# CONSTRUCTION
# MATERIALS

Prentice-Hall, Inc.     *Englewood Cliffs, New Jersey*

**W. J. PATTON**

*Red River Community College*

# CONSTRUCTION MATERIALS

*Library of Congress Cataloging in Publication Data*

PATTON, W J
  Construction materials.

  1. Building materials.  I. Title.
TA403.P318      691      74–28119
ISBN  0–13–168724–7

© 1976 by
Prentice-Hall, Inc.
Englewood Cliffs, N.J. 07632

10  9  8  7  6  5  4  3  2  1

Printed in the United States of America

PRENTICE-HALL INTERNATIONAL, Inc., London
PRENTICE-HALL OF AUSTRALIA, Pty Ltd., Sidney
PRENTICE-HALL OF CANADA, Ltd., Toronto
PRENTICE-HALL OF INDIA PRIVATE LIMITED, New Delhi
PRENTICE-HALL OF JAPAN, Inc., Tokyo

# CONTENTS

# 2 THERMAL CHARACTERISTICS 24

# 3 ACOUSTICS AND CONSTRUCTION MATERIALS 35

*Part* **2** THE MATERIALS

# 4 ROCK AND SOIL 47

**APPENDICES**

There are few careers more interesting and lively than those associated with construction. The construction materials are as interesting as the jobs, especially so when you have mastered their technology and can exploit them in innovative and imaginative ways. New ones continually appear, bringing with them new application techniques, and new ways for the imprudent or uninformed architect, applicator, specifier, or estimator to make those embarrassing errors that are an inevitable part of the construction scene.

Construction materials and methods are much more sophisticated than they sometimes appear, and a systematic and fundamental approach is required to understand what is possible and what is not possible with them. The first three chapters undertake an examination of basic properties, including an introduction to that vexing problem, the fire resistance of materials. Part 2 examines the established materials and also those newer types for which there appears to be sufficient building experience to assess their dependability. Part 3 then reexamines these materials in terms of the service required of them in wall, roof, and floor applications.

There is, of course, no substitute for experience and experiment in finding the potential and shortcomings of building materials. To encourage practical tests on the part of the reader, pertinent ASTM (American Society

# *AUTHOR'S*
# *REMARKS*

for Testing and Materials) specifications are given at the ends of the chapters, and other investigations of a less formal nature are suggested.

For those using the book for formal courses in colleges and institutes, a glossary of construction terms is provided at the end of the book.

A word of caution regarding items that could be handled only in the briefest fashion. Chapter 4 deals with two very comprehensive and specialized subjects. To discuss both rock and soil fully would require at least half the pages of this book—and that merely for an introduction. This chapter therefore cannot serve as a complete presentation of soil and rock studies. Also, although the use of construction materials is strictly governed by building codes, the complexities (which include politics) of these codes could not be discussed here beyond acknowledging that they exist and must be consulted.

I trust that architects, designers, builders, and materials salesmen will find fresh and stimulating points of view here, as well as information and an organized view of the whole subject of construction materials and their competitive merits.

W. J. P.

*CONSTRUCTION
MATERIALS*

# PROPERTIES
# OF CONSTRUCTION
# MATERIALS

*Part*

## 1.1  three fundamental groups
## of construction materials

With few exceptions, construction materials are solid materials or harden into solid materials. Solid materials are grouped into three fundamental types: ceramics, metals, and organics.

The *ceramic* materials are rock or clay minerals, or are compounded from such minerals. Examples are sand, limestone, glass, brick, cement, gypsum, plaster, mortar, and mineral wool insulation. These are materials dug from the earth's crust with or without further processing and purification. Since they are extracted from the earth, they are relatively inexpensive as compared to metals or the organic materials. The ceramics have been used as building materials from time immemorial, and their virtues will ensure their use in the future: they are enduring, hard, and rigid. Their outstanding disadvantages are brittleness and their heavy weight.

*Metals* are extracted from natural ores, which of course are also ceramic materials. Such metallic ores are usually oxides or sulfides of metals. The metals are not as hard as the ceramic materials and, because they must be

# CHARACTERISTICS
# OF CONSTRUCTION
# MATERIALS

# 1

extracted from the ore by complex smelting processes, they are more expensive. Ceramic materials are brittle, and so are restricted to the carrying of compressive forces in buildings and structures. Metals are ductile and are used where tensile forces must be carried.

Ceramic materials do not corrode in the atmospheric conditions to which buildings are exposed; metals do. The corrosion process returns the metal to its original state as a mineral. When iron or steel rust, they oxidize to iron oxide, $Fe_2O_3$, which is the iron ore hematite. Aluminum oxidizes to $Al_2O_3$, which is the ore bauxite.

The *organic* materials are largely a development of the twentieth century, with the notable exception of wood and bitumens. These are the numerous and increasing synthetic materials based chemically upon carbon. The organics include wood, paper, asphalts, plastics, and rubbers.


## 1.2   introduction to

## the materials

Of all the available building materials, *rock* and *stone* are traditionally associated with permanence. These are the enduring materials of the earth's crust, the oldest being perhaps 4 billion years old. But there is a remarkable difference in service requirements between a relatively static and protected position in the earth's crust and a building wall, and no rock will effectively last forever when incorporated into a building. All rocks are porous to some degree and can absorb water, with all the possibilities of water for damage and deterioration.

*Concrete* is a man-made conglomerate rock material of stone aggregate, sand, and a cement adhesive. It has a more complex structure and chemistry than rock and hence is exposed to a greater variety of modes of deterioration. It is weaker than rock in tension and compression and is not expected to support tensile forces. Like any other composite material, it is only as strong as its bonding material, the cement. Finally, it must be emphasized that the quality of concrete is dependent on the skill with which it is compounded and handled on the job site.

The architect's problem with concrete as a building facing material is that of satisfactory appearance. Its color is drab gray. But as a construction material it offers substantial advantages:

1. Concrete may be formed into virtually any shape.
2. It is both hard and rigid.
3. It is inexpensive.

*Brick* is a burned ceramic product somewhat similar in chemistry to concrete and cement. It is made in modular sizes in a range of colors. It is more attractive in color, texture, and general appearance than concrete, and quite elaborate designs are possible in brick.

The outstanding characteristic of *glass* is its transmission of light and other radiation.

*Metals* offer the advantages of high strength in compression, tension, or shear, ductility, rigidity, hardness, and dimensional stability. Their susceptibility to corrosion has been mentioned, although certain alloys such as the stainless steels are not corroded by the service conditions found in buildings.

The American architect Frank Lloyd Wright expressed the appeal of *wood* in the following words: "Wood is universally beautiful to man. It is the most humanly intimate of all materials. Man loves his association with it, likes to feel it under his hand." Wood is highly corrosion-resistant and remarkably durable, weathering away at a rate of about $\frac{1}{4}$ inch per century. Although plastics and metals have increasingly displaced wood in building products and manufactured products, it is not conceivable that wood will become obsolete because of its unique advantages over other materials:

1. Trees are a self-perpetuating natural resource.
2. Wood is, or can be made, relatively abundant in most geographical areas.
3. Wood can be shaped with ease and with the use of simple tools.
4. Wood, like all organic materials, is light in weight.
5. It is perhaps the most beautiful of construction materials, with a variety of natural designs and textures.
6. It is the strongest and until recently the cheapest of the cellular construction materials.

Wood, clay, and stone have been standard construction materials for at least 100,000 years. Brick and concrete have a history of thousands of years. By comparison, the *plastics* and *rubbers* are recent additions to the range of construction materials. They have been serious contenders in construction for only about 20 years. They are a special case, and call for some special remarks.

## 1.3 organic construction materials

A technical man's view of history would probably correspond to that of Fig. 1.1. The decade 1850–1860 includes the major event in materials technology: in this decade new developments in steel-melting practice made

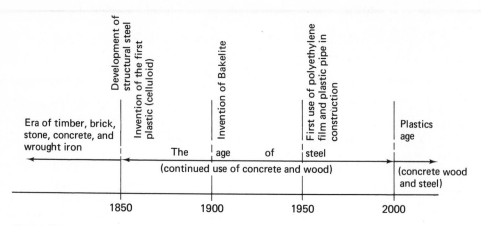

Fig. 1.1 Technical man's view of history.

structural steel cheap and its use widespread. Before this time, construction materials were largely confined to ceramic materials and wood, with a limited use of cast iron and wrought iron, especially for bridges. The Steel Age in which we live thus is not much more than 100 years old. Steel is the dominant material in manufacturing generally, being consumed in North America at the rate of about 150,000,000 tons per year.

But a study of trends in materials usage and of price trends of materials, together with an assessment of developments in that most versatile technology, plastics science, suggests some dramatic changes in materials usage in the near future. It appears to be clearly indicated that in or before the year 2000 the tonnage of plastics consumed will exceed that of steels. Indeed, it is already true that the number of cubic inches of plastics annually consumed exceeds that of steels. The Steel Age therefore will represent a comparatively brief historical interval of not more than 150 years, so this outstanding technical age is close to its end. It would seem that for the indefinite future, plastics and rubbers will assume a dominant position in the materials of construction and manufacture.

However, it would be unwise to assume that plastics and rubbers will make other construction materials obsolete. Every material excels in certain characteristics, yet every material has its disadvantages.

In addition to the plastics and rubbers, the organic group includes many other materials of significant importance to architecture and construction: wood, paper, asphalts, sealants, surface finishes, adhesives, solvents, oils, and explosives.

If we compare the structural organic materials, which are wood, rubber, and plastics, with the structural metals and ceramics, we find a number of differences:

1. The organics are softer and have lower strengths.

2. They lack stiffness, and under load may deflect excessively.
3. With some exceptions, most organic materials have unusually high ductility, especially in tension. Some rubbers may be elongated to as much as ten times their original length under stress.
4. The organic materials cannot carry significant loads at temperatures much above ambient temperatures.
5. The organic materials are the lightest in weight of all the construction materials.
6. Thermal conductivities are low for plastics, rubbers, and wood, and all are valuable as heat insulation.
7. Some of these materials are combustible but can be made flame-retardant.
8. They offer attractive color effects.
9. They are available in a wide range of substances and formulations (including wood), and new formulations appear every year.

Since all construction plastics are relatively new in the construction field, and since new ones will be constantly offered to the market, architects and builders will never be as familiar with these materials as with the more traditional construction materials.

## 1.4 deterioration

## of building materials

The function of a building includes the maintenance of a satisfactory internal environment and spatial enclosure, as well as that of an acceptable appearance. These functions must be supported over a period of decades. But all materials deteriorate through time, although at varying rates. Much of the deterioration is due to weathering, an aging process resulting from the varied destructive effects of the environment and climate. Other deterioration arises from the effects of occupancy of the building: floors and stairs become worn, doors and door frames are damaged by repeated use of the doorway, and walls are damaged by contact with furniture.

## 1.5 temperature variation

Temperature changes are the cause of expansion and contraction in building materials. This is not always a serious effect except when there is differential expansion due to differences in coefficient of thermal expansion. The aluminum convector cover of Fig. 1.2 is fastened to a slate board. It is scarcely possible to find two materials with a wider difference in coefficient of expansion—of the order of 5:1—hence the buckling of the aluminum, which has the higher thermal expansion. Similarly, aluminum roof flashings have a higher expansion than the asphalt of a roof deck, and the design of the roof

Fig. 1.2 Buckling of an aluminum convector cover due to thermal expansion.

must take account of this difference. The following list gives the amount of thermal expansion in inches per 100-ft length for a temperature change of 100°F:

| brick | 0.36 | steel | 0.72 |
|---------|------|----------|------|
| glass | 0.54 | aluminum | 1.46 |
| concrete | 0.66 | vinyl | 3.6 |

The similar thermal expansions of steel and concrete must be noted. This makes it possible to reinforce concrete with steel rods. Aluminum-reinforced concrete would be possible only if temperature variations were insignificant and of course if concrete did not corrode aluminum.

## 1.6  movement

It is the business of the mechanical engineer to design equipment that moves. The layman assumes that buildings, bridges, roads, and other monumental structures of the civil engineering and architectural fields do not move.

But the architect, the civil engineer, and the builder all know that structures do move. The Leaning Tower of Pisa does not lean because the builder forgot to plumb it. An occasional structure has even fallen on its side. Plaster, brickwork, and concrete crack, floors heave, and roofs leak. Concrete shrinks and wood swells or contracts as its moisture content changes. All these failures are movements. Some of these movements arise from overloading or instability of soil fundations (see Fig. 1.3); others arise from thermal expansion; still others have more complex causes. Many of these structural movements cause no risk of building collapse, but may result in further movement or damage by infiltration of water or frost action.

Often it is not possible to prevent these building movements. Instead, the building designer tries to assess the degree of movement likely to occur, and he allows for this in his design. The amount of movement that can be accommodated will depend on the rigidity of the structure. A timber building may tolerate formidable movements and still stand, while a brick or monolithic concrete structure will be less tolerant of such building movements.

Movements due to changes of moisture content can affect only absorbent materials such as clay, wood, nylon, or some types of plaster. Worse damage may occur if water collects in cracks or openings and then freezes. The expansion of water on freezing is a volume increase of almost 10 percent. Such moisture penetration and frost action is especially harmful to roofs and roof insulation.

Fig. 1.3 Cracks in a concrete block wall produced by soil shrinkage.

Soil movements are a major subject of study. Sands and gravels do not absorb water and are not subject to expansion or contraction with changes of moisture content. But the very fine particles of which clay consists are highly absorbent, so clay swells and shrinks remarkably with variations of moisture content.

Even the corrosion of steel results in harmful expansion effects. If steel should rust, the rust is a porous material that readily takes up moisture. Expansion is due both to the increased volume of the rust and its absorbed water. If the steel is buried in other materials, these may be cracked or spalled off. Such cracking and spalling may occur if reinforcing rods are too close to the surface of the concrete in which they are embedded.

## 1.7 oxidation

Paints, varnishes, rubbers, sealants, asphalts, and many (but not all) plastic materials will oxidize slowly with time. Oxidation makes the material harder and more brittle, with the final result that the material may develop cracks, crazing, or deep fissures. An oxidized asphalt roof cannot "give" in response to building movements and expansion; and oxidized sealant loses the flexibility required of a seal.

## 1.8 strength and ductility

The strength of materials is the critical concern in the structure and frame of a building. Various strength requirements also influence the choice of construction materials in other building components; for example, a floor must not indent under loads and a roofing material must not sag. Materials must uphold the stresses imposed by reason of the normal occupancy of the building.

Not strength, but ductility, may be required of other building components. Roofing materials must be able to "give" with movements of the building. A sealant must be able to expand and contract to keep closed the joint that it seals. An adhesive must accommodate thermal movements in the materials that it joins.

Hence, for any structural material, the properties of stress and strain must be known with reasonable accuracy, including any changes in these properties that may occur with time and exposure. Thus if a specimen of a material shows an ultimate tensile strength of 10,000 pounds per square inch (psi), the designer must be able to assume that all shipments of the same grade of material will possess at least that tensile strength. This predictability certainly holds for manufactured materials such as metals and plastics. But it is less certain for wood, unless the wood is carefully selected and graded, and quite uncertain for stone, which may vary in quality in different areas of a quarry and also in the depth from which it was quarried.

While the words "stress" and "strain" are often confused in ordinary conversation, each has a specific meaning. Consider the case of a 4 × 4 ft timber post 100 in. long supporting a compressive load of 400 lb (Fig. 1.4). The *stress* in the post is the load divided by the cross-sectional area over which the load is distributed:

$$\text{stress} = \frac{\text{load}}{\text{area}} = 400\text{lb}/16 \text{ in.}^2 = 25 \text{ psi}$$

There are only three possible types of stresses: *tension, compression,* and *shear*. A shear force or stress tends to cut through the material, as when a hole is punched in a metal sheet, or it may be a twisting force, as exerted by a screwdriver or a wrench. The strength of materials is lower in shear than in tension or compression, and for most materials, especially brittle materials such as concrete, the strength in tension is lower than in compression.

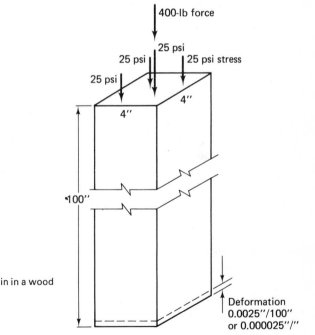

Fig. 1.4  Stress and strain in a wood post.

Accompanying any stress, there is always some deformation (change in dimension) of the loaded member. In the present example the post will shorten in length under the load. Suppose that the length of the post under load decreases by 0.0025 in. The amount of this deformation will depend on the length of the post: if the post were twice as long, it would shorten twice as much under the same load or stress. It is therefore convenient to express this deformation as deformation per inch of length. Deformation per inch is the *strain* or *unit strain*, which in the present example is 0.0025 in./100 in. = 0.000025 in./in.

Materials may be strained by conditions other than stress. If the temperature of a material is increased, it will expand. A thermal strain is supposedly recoverable when the temperature returns to the original level of temperature, but such is not always the case, as in Fig. 1.2.

If the stress is removed from a material, two types of strain behavior are possible:

1. The material may return to its initial size and shape. This type of behavior, in which strain disappears with stress, is termed *elasticity*. Typical elastic materials are glass, Bakelite, and rubbers. Brittle materials are necessarily elastic.
2. The material may remain permanently deformed by the stress. This type of strain behavior is termed *plasticity*, familiar in moist clay, uncured concrete, and polyethylene film.

Many materials are elastic under conditions of low stress, but plastic at high stress levels. The word "ductility" implies an ability of the material to strain plastically: polyethylene film is ductile, but glass is not.

# 1.10   *stress-strain diagrams*

The stress-strain diagram of a material is a most useful compilation of information about any material, because it discloses in a single diagram such characteristics as ultimate strength, limit of elastic behavior, ductility or brittleness, elasticity, plasticity, and modulus of elasticity.

A stress-strain diagram may be given for compression, tension, or shear. The most common is the tensile stress-strain diagram, although compression tests may be more useful for those materials that are normally loaded in compression, such as concrete or plastic foam insulations.

Figure 1.5 is a tensile stress-strain diagram for a mild (i.e., soft) construction steel. On being loaded, the steel specimen at first strains linearly. This linear behavior is usually an indication of elasticity. At about 35,000 psi the strain appears to become plastic, since the material deforms at an almost constant load. At about 4 percent strain the material takes up more load. The maximum or ultimate tensile stress is 67,000 psi, and thereafter the load falls as the specimen necks down to a smaller diameter.

The elastic limit or limit of elastic strain is presumably somewhere in the

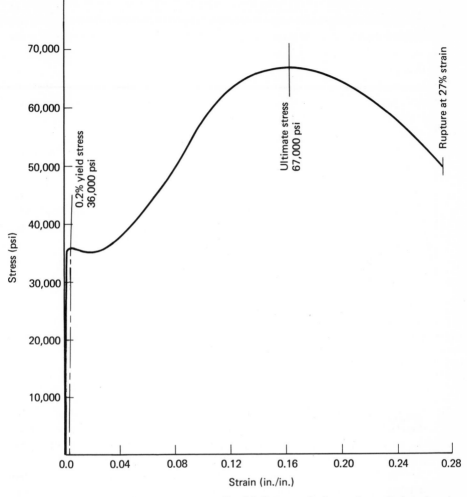

Fig. 1.5 Stress-strain diagram for a rolled-steel section.

region of 36,000 psi for this specimen; it is not necessarily the terminal point of the linear portion of the stress-strain curve. Because the elastic limit is somewhat indeterminate, the 0.2 percent yield stress is substituted for it. This yield stress is determined from the stress-strain curve by drawing a line from the strain axis at 0.002 strain, parallel to the elastic portion of the graph. The yield stress is the point on the stress-strain curve intersected by this line. For this steel specimen, the yield stress is 36,000 psi.

A more complex example of stress-strain behavior is given in Fig. 1.6. This is a compression stress-strain curve for an insulating grade of rigid foamed polyurethane with a density of 4 lb per ft$^3$. On being loaded, the material at first strains linearly like the steel specimen. This linear strain is

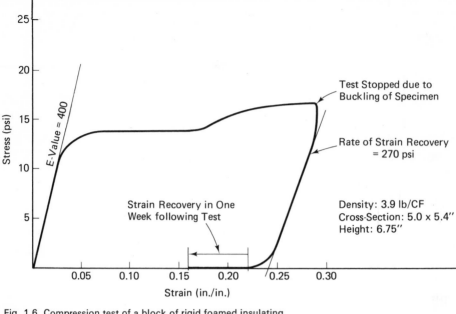

Fig. 1.6 Compression test of a block of rigid foamed insulating polyurethane 3.9 lb per ft³. Almost one-half of the total strain was recovered in 1 week, indicating that this "plastic" is actually a rubber. Note the different slopes of the linear loading and unloading lines.

again elastic. At about 12 psi, the strain seems to become plastic, since the material deforms continually at a constant load. At about 18 percent strain the material takes up more load. The loading was stopped at about 17 psi, owing to incipient buckling of the block of urethane.

On being unloaded, the material recovers elastically at approximately the same rate shown during the initial loading. This elastic recovery reduces the strain from the maximum of 29 percent to 22 percent. The curved portion of the strain-recovery line near zero stress is the recovery during 15 minutes after the test. After 1 week, the permanent strain fell to 16 percent. Thus approximately half the maximum strain is elastic and half plastic, and the elastic limit must be far beyond the linear portion of the stress-strain curve. Note especially that the elastic strain recovery was not instantaneous with removal of the load. Strain is not proportional to stress except in the case of certain materials, usually metals, at low stress levels and slow loading.

The end of the linear portion of the stress-strain diagram is termed the *proportional limit.* For the polyurethane foam of the figures, the proportional limit is 12 psi. The yield stress is a more useful figure than the proportional limit as the assumed limit of elastic behavior in metals. On the other hand, the yield stress for organic materials is a rather vague concept, and the proportional limit is used instead. For these materials, the proportional limit is simply the limit of the linear relationship between stress and strain; it is not necessarily the limit of the elastic range. It is not, for example, the limit of the elastic range in the case of the polyurethane foam.

A material such as this polyurethane, with a maximum elongation of about 30 percent, has excellent ductility. It could expand or contract to accommodate movements or settlement of the building in which it is installed. It could serve, for example, as a sealant. On the other hand, this material is quite obviously unsuited to the support of building loads: it has little strength and deforms excessively. In the materials of any building, some must be strong and stiff to support the building loads; others must be flexible to accommodate building movements.

## 1.11   modulus of elasticity

Most metals, some rubbers such as urethanes, and the elastic plastics such as Bakelite and reinforced plastics, exhibit linear elastic behavior below a certain stress termed the elastic limit (the urethane of Fig. 1.6 has an apparent, but not real, elastic limit of 11 psi). The elastic limit is the highest stress for which the strain is elastic. The real elastic limit of the urethane under discussion must be 14 psi, since some of the horizontal portion of the graph at this stress must be elastic and some plastic.

If the initial portion of the stress-strain graph is linear, that is, strain is proportional to stress, then the material has a modulus of elasticity. *Modulus of elasticity*, or *Young's modulus*, is defined as the ratio of stress to strain:

$$E = \frac{\text{increase in stress}}{\text{increase in strain}}$$

More informally, modulus of elasticity is referred to as *E value*, with the symbol *E*. The modulus of elasticity may also be considered as the stress necessary to strain the material elastically to twice its original length, if that were possible.

A very rigid material, such as steel, had a very high *E* value, since a large stress will be accompanied by only a small strain. *E* values for metals and many ceramic materials are greater than $10 \times 10^6$. Most plastics have *E* values in the range of a few 100,000, indicating that they are less rigid, or yield more readily under stress, than metals.

Since the *E* value is the ratio of stress to elastic strain, it is given by the slope of the initial stress-strain curve. This is drawn in Fig. 1.6, showing a modulus of elasticity for the urethane insulation of 400.

| | | | |
|---|---|---|---|
| glass | $10 \times 10^6$ (varies) | polyethylene | $35 \times 10^3$ |
| concrete | $3 \times 10^6$ (varies) | Bakelite | $1.1 \times 10^6$ |
| wood | $1.5 \times 10^6$ (varies) | reinforced plastics | $1 \times 10^6$ (varies) |
| steels | $29 \times 10^6$ | rigid vinyl | $0.35 \times 10^6$ |
| aluminums | $10 \times 10^6$ | | |

This table of $E$ values suggests that the figure of $1 \times 10^6$ separates the high-modulus or stiff construction materials from the low-modulus plastics and rubbers. The high-modulus steels and concretes must be used for the columns, beams, and slabs of conventional building frames, since these materials can carry large loads with minimal deflection. There is no present possibility of high-modulus plastic materials being developed, and these materials cannot be substituted in building frames.

## 1.12   *modulus of rupture*

Certain very brittle construction materials such as brick are difficult to test in tension. For such materials, the modulus of rupture, rather than the tensile strength, is determined. The *modulus of rupture* gives the tensile strength at failure in a bending test. It is unrelated to the modulus of elasticity.

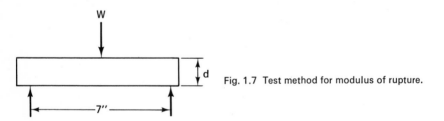

Fig. 1.7 Test method for modulus of rupture.

In a test for modulus of rupture, the material is loaded as a beam as shown in Fig. 1.7. From the test results, the modulus of rupture is calculated by the following formula:

$$\text{modulus of rupture, psi} = R = \frac{3WL}{2bd^2}$$

where $W$ = load at failure, pounds
$L$ = loading span, actually 7 in. in the standard ASTM tests
$b$ = width of specimen, inches
$d$ = depth of specimen, inches

The stress-strain curve of any such brittle materials is a straight line to the failure stress. Brittle materials are elastic materials, although their elasticity is extremely limited.

## 1.13  architectural shapes
## and material strength

The theory of strength of materials shows that the resistance to buckling of struts and columns (Fig. 1.8) is proportional to the modulus of elasticity of the column. If plastics are to be used as load-carrying members of a building frame, they cannot be loaded in compression because of poor resistance to

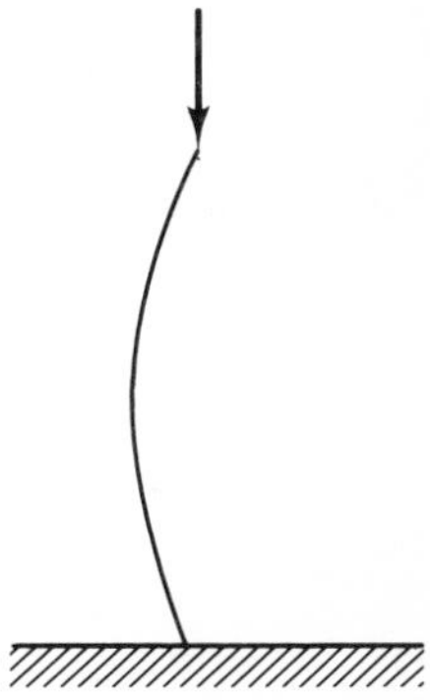

Fig. 1.8 Thin columns fail in buckling, not in compression.

buckling. They are successfully used, however, in those structural forms that place the plastic material in tension or in which the structural shape of the building provides the required stiffness. Alternatively, these low-modulus materials may be incorporated into sandwich materials, a type of material that is to be discussed presently. Some of the types of building shape suitable for plastic materials are shown in Fig. 1.9. The advent of construction plastics has stimulated architects and structural engineers to produce some remarkably imaginative building shapes.

Steel and timber provide adequate tensile strength and can support bending stresses (Fig. 1.7). Concrete is strong in compression but has little tensile strength. If reinforced with steel on the tension side of a concrete beam, concrete is entirely suitable as a material in flexure.

Brick and stone are suited only to the support of compressive forces. They are used in the construction of arches, vaults, or spherical domed roofs, where the stress is wholly compressive.

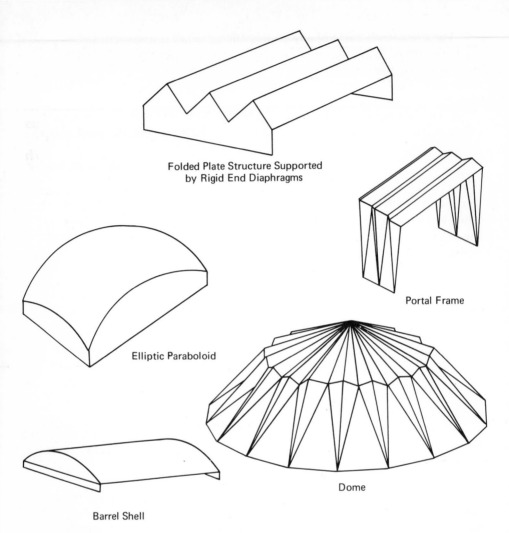

Folded Plate Structure Supported
by Rigid End Diaphragms

Portal Frame

Elliptic Paraboloid

Dome

Barrel Shell

Fig. 1.9 Selection of building shapes suited to construction in
plastics. A metal building frame is usually necessary.

## 1.14 factor of safety

The relative safety of the structural component is usually given in the *factor
of safety*. This is the ratio of the maximum stress in the component to the
maximum stress the material can support without failure. The more predict-
able the loads and the more consistent the quality of the material, the smaller
this factor of safety can be; in the ideal case the factor of safety would be
required to cover only unforeseen loads.

The meaning of failure in the factor of safety must be clearly understood.
Failure may be crushing of the material in compression, tearing in tension, or

splitting in shear. But uncontrollable deformation must also be considered failure of the material. For ductile materials such as construction steels, uncontrollable deformation begins at the yield stress, well before failure in tension or compression, and it is usual to relate the factor of safety to this yield stress rather than ultimate stress.

Consider a steel beam with a yield stress of 36,000 psi in tension and an ultimate tensile stress at rupture of 67,000 psi. Suppose that the beam is allowed a maximum stress in flexure of 22,000 psi. Then its factor of safety is

$$\frac{22,000}{36,000} \quad \text{or approximately 1.65}$$

It is scarcely realistic to give the factor of safety as 3.0 based on the ultimate stress.

## 1.15   creep

Certain materials are subject to creep, especially the plastics used in construction, and to a lesser degree wood and concrete. *Creep* is the slow but constant increase in deformation under a permanent load, and is a plastic deformation. Creep is not a characteristic of steels or aluminums as used in construction.

Provided that stresses are held below a certain limit, creep will be negligible. In the case of concrete, this creep limit is higher than the stresses allowed in concrete construction codes. While the effects of creep may be found in wood construction occasionally, they are a very basic consideration in plastics.

The allowable stress in a plastic material is limited by the amount of creep that can be allowed, rather than by the considerably higher stresses that short-term loading would allow. This creep limit may be 40 percent of the rupture stress in the more durable construction plastics. If the allowable stress is held at a level slightly below the creep limit, as is commonly done, the factor of safety is little better than 1.0. Such a factor of safety suggests foolhardiness in design. But such a design concept is quite safe, since the plastic material can support temporary stresses well in excess of the creep limit. Actually, the method of designing by factor of safety is not applicable to plastic materials.

## 1.16   sandwich construction

Both traditional and newer types of construction materials and components use composite construction, either laminated or sandwich types. A familiar example is the built-up roof, which is a complex laminate of alternating layers of roofing felts and asphalts. Plywood is a laminate. Glued laminated structural timbers are popular for the framing of churches and arenas. Even paint finishes are laminated. The effectiveness of such composite materials lies not in the components, but the complete laminate—the whole is greater than the sum of the parts. Therefore, such materials are only as good as their adhesion. It would be quite useless, for example, to test the effectiveness of a detached film of paint. The paint film must be attached to the substrate—wood, metal, concrete, or other—which it is to cover and protect, and the effectiveness of the paint is the composite effectiveness of the paint and its bond to the substrate.

Figure 1.10 shows an example of sandwich construction, in this case an experimental type with both core and faces of rigid foamed polyurethane. The core is the same polyurethane insulation as in Fig. 1.6 between faces of hard polyurethane of density 22 lb per ft³. The faces of the sandwich must

Fig. 1.10 Polyurethane sandwich using lightweight insulating polyurethane core and faces of polyurethane 22 lb per ft³. The faces are sprayed against the core without adhesive.

provide the required load-bearing strength, stiffness, hardness, and surface characteristics. The core is soft and yielding and so is not used to carry the direct structural loads. Instead, the function of the core is to keep the faces apart and to stabilize them against buckling when loaded. Frequently, however, the core is given additional functions such as heat insulation. The effectiveness of a sandwich construction may be understood by considering the case of a structural wall made of two faces of thin aluminum sheet with a core of foamed polystyrene or polyurethane. Without the support of the core, the thin aluminum sheets would buckle under a load of a few pounds. Again, the core by itself is too weak to carry any significant load. But if the two materials are bonded into a sandwich, the strength of the sandwich is remarkably high.

A stress-strain diagram for the sandwich of Fig. 1.10 is given in Fig. 1.11. Note the initial linear reaction between stress and strain. The maximum stress was 100.9 psi, at which stress one face began to delaminate from the

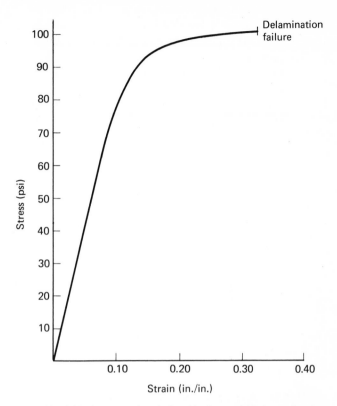

Fig. 1.11 Stress-strain relationship for a rigid foamed polyurethane sandwich in compression. Core density: 4 lb per ft³ 1
in. thick faces: 0.15 in. thick, density 22 lb/ft³.

core. No adhesive was used to join the faces to the core; instead, the face
foam was simply sprayed to the core foam. Comparing Figs. 1.6 and 1.11,
the presence of thin faces of a slightly harder and stronger material raised the
ultimate compressive strength by a factor of 6. The stress in the sandwich is
averaged over the whole cross section of the sandwich even though the load
is taken on the faces. Much higher stress would be possible if a material of
higher $E$ value were used for the faces, such as aluminum or steel. Sandwich
materials are a development that is still progressing. This type of structure is
still less used in construction than in aircraft, motor homes, shipping containers, and temporary housing, but its advantages suggest that it will be
widely used in future construction. Among its advantages are these:

1. Weight savings, since the core can be lightweight and the faces thin.
2. Suitable surface finish and resistance to deformation and indentation.
3. Rigidity.
4. Thermal insulation.
5. Ease of production and installation.

Pertinent specifications of the American Society for Testing and Materials (ASTM) follow, and will be listed at the end of each subsequent chapter.

## PERTINENT ASTM SPECIFICATIONS

E8–69 (Part 6) Tension Testing of Metallic Materials
E251–67 (Part 30 or 31) Performance Characteristics of Bonded Resistance Strain Gages
E328–67T (Part 30 or 31) Stress-Relaxation Tests for Materials
E9–67 (Part 31) Compression Testing of Metallic Materials
E16–64 (Part 31) Free Bend Test for Ductility of Welds
E111–61 (Part 31) Determination of Young's Modulus at Room Temperature
A370–71b (Part 3,4, or 31) Mechanical Testing of Steel Products
C393–62 (part 16) Flatwise Flexure Test of Flat Sandwich Construction
C480–62 (Part 16) Flexure-Creep of Sandwich Constructions
C165–54 (Part 14) Compressive Strength of Preformed Block-Type Insulation
E72–68 (Part 14) Strength Tests of Panels for Building Construction
E6-66 (Part 31) Definitions Relating to Mechanical Testing

## INVESTIGATIONS

1  Test several varieties of common and face brick for modulus of rupture.

2  In accord with ASTM E9-67, make compression stress-strain tests on polyurethane foam and foamed polystyrene. Determine the modulus of elasticity in compression.

3  In accord with ASTM E8-69, make tension stress-strain tests on the following materials. Compare and explain the differences between the several stress-strain curves. For any linear stress-strain relationship, determine the modulus of elasticity. Suggested materials: a construction steel, a stainless steel alloy 302 or 304, a reinforced fiberglass plastic sheet, polycarbonate, polyethylene, a rubber of the hardness of an automobile tire or a gasket rubber.

## QUESTIONS

1  Differentiate among proportional limit, elastic limit, and yield stress of a metal.

2  What is the difference between modulus of elasticity and modulus of rupture?

3  Define factor of safety.

**4** Define creep.

**5** A specimen of material 2 in. wide and $\frac{1}{2}$ in. deep is tested for modulus of rupture over a span of 7 in. The specimen cracks under a load of 870 lb. Determine the modulus of rupture.

**6** An aluminum flashing is installed against a brick parapet wall. For a temperature increase of 75°F, how much longer does the flashing grow than the brick wall?

### 2.1 fire hazard

It might be thought that the matter of fire hazard is a simple enough one: does the material burn or not? Wood burns. Steel does not. While these statements about wood and steel are certainly true, they are almost irrelevant to the relative fire risk of the two materials. Compare the cases of fire in two different buildings: one framed of heavy timbers such as glued laminated wood arches, and the other of light steel frames and roof joists. The steel roof joists will collapse after a relatively few minutes of exposure to fire, while it may require several days of fire to bring down the timber framing. The subject of fire hazard turns out to be a complicated one.

Fire characteristics of building materials are defined in the answers to the following questions:

1. Is the material combustible?
2. If combustible, does it ignite readily and spread the flame rapidly?
3. What are its burning characteristics?
4. Finally, if it does not burn, how is it influenced by adjacent burning materials?

# THERMAL
# CHARACTERISTICS

2

Some materials may burn quite slowly but may propagate a flame very rapidly over their surfaces. Thin wood paneling will burn readily, yet a heavy timber post will sustain a fire on its surface until charred and thereafter smoulder at a remarkably slow rate of burning. Bituminous materials may spread a fire by softening and running down a wall. Steel does not burn but is catastrophically weakened by the elevated temperatures of a fire. Glass is shattered by heat. Fire hazard is not, therefore, a subject for which there are easy answers and conclusions.

Protective treatments are possible for all combustible materials. Fire-retardant grades of foamed plastics are available, and fire-retardant paints offer further protection.

### FIRE HAZARD CHARACTERISTICS

| Characteristic | Styrofoam | Urethane Foam | Wood |
|---|---|---|---|
| heat of combustion, Btu/board foot | 2700 | 1800 | 26,600 |
| self-ignition temperature | 725–900 | 975 | 500 |
| ASTM E84 Tunnel Test for Surface Flame Spread | 10–25 | 25–80 | 100 |

Foams classified as *self-extinguishing* when tested in accord with ASTM D-1692 must extinguish themselves within 10 seconds. When exposed to a flame, foamed polystyrene tends to melt away and thus extinguish itself. Foamed polyurethane tends to char and maintain its shape and size. The low values for heat of combustion of foams in the table result from the low densities of the foamed plastics. ASTM E84-68, Surface Burning Characteristics of Building Materials, is one of the more important fire hazard tests, and should be consulted.

## 2.2  thermal conductivity

In order for heat in a room to be conducted through a material such as a building wall, the heat must pass through three barriers. First it must get past the surface of the material, then pass through the material to its opposite surface, and finally the heat must be transferred from the building wall to the outside air. There are two surfaces passing heat to or from the surrounding air.

Consider first the transfer of heat through the material of the wall. The thermal conductivity of a material is usually referred to as its $K$ factor. A low $K$ factor, below 1.0, indicates that the material is a heat insulator. Metals conduct heat readily and have $K$ factors from about 100 to 2700.

The unit of heat currently used in the building industry is the *British thermal unit* (*Btu*), which is the amount of heat required to raise the temperature of 1 lb of water by 1°F. The hour is the usual unit of time for heat calculations, hence heat flow is given as Btu/hr, often compressed to Btuh.

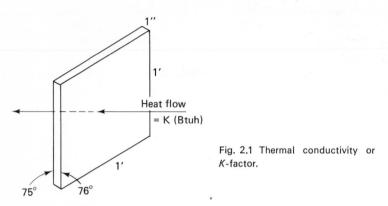

Fig. 2.1 Thermal conductivity or $K$-factor.

Figure 2.1 shows a board foot of an insulating material, styrofoam or beadboard (foamed polystyrene). This material has a $K$ *factor* of 0.24. Suppose that there is a temperature difference of 1°F between the two faces of the styrofoam. Then heat will flow from the hot face to the cooler face. The actual heat flow in Btuh through a board foot of styrofoam with a face difference of 1°F will be 0.24 Btuh. The $K$ factor of any material therefore is the heat flow or heat loss per hour through a board foot of the material at a temperature difference of 1°F. Since the $K$ factor is based on a board foot, it has mixed units of feet and inches:

$$K \text{ factor} = \text{Btuh/ft}^2 - °\text{F-in. thickness}$$

The heat transfer through a wall area will be proportional to the area and also to the temperature difference between one face and the other. It will be inversely proportional to the thickness of the insulation:

$$\text{heat flow through a material} = Q = \frac{KA \, \Delta t}{L}$$

where $K = K$ factor or thermal conductivity, Btuh/ft$^2$ — °F-in.
$A$ = area of surface, ft$^2$
$L$ = thickness of the material, inches
$\Delta t$ = temperature difference across the material, °F

**Example.** Fiberglass insulation has a $K$ factor of 0.3. What is the amount of heat conducted per hour through 24 ft$^2$ of fiberglass insulation 3 in. thick, if the hot face temperature is 100°F and the cold face 35°F?

$$Q = \frac{KA \, \Delta t}{L} = \frac{0.3 \times 24 \times 65}{3} = 156 \text{ Btuh}$$

The *conductance* $C$ of a material or a system of materials is the number of Btuh conducted through 1 ft² of the material or materials for a 1°F temperature difference between the faces. $C = K/L$, where $L =$ material thickness. For example, $C$ for haydite concrete (concrete made of expanded clay aggregate) 3 in. thick is 0.22, indicating that 1 ft² of this material conducts 0.22 Btuh for a 1°F temperature difference.

Frequently, the $R$ *factor*, or resistance to heat flow, is used. This is the reciprocal of $K$ or $C$. Thus fiberglass insulation has a $K$ factor of 0.3, or an $R$ factor of 3.3. Three inches of haydite concrete has a $C$ of 0.22, or an $R$ factor of 4.54.

In the case of a composite wall, $U$, the overall coefficient of heat transmission (see Fig. 2.2), is the number of Btuh conducted per square foot of surface per °F temperature difference across the wall. To determine the $U$ *factor*, the $R$ values of the several materials are added, and $U$ is the reciprocal of the total $R$. This procedure can be clarified by an example.

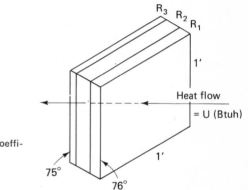

Fig. 2.2 *U*-factor (overall coefficient of heat transmission).

A cold storage ceiling is built of $\frac{3}{4}$-in. plywood, $R = 0.59$, and 4 in. of sprayed polyurethane insulation, $K = 0.16$. Find the $U$ factor for this construction.

The $R$ factor for 4 in. of urethane $= 1/0.04 = 25$:

|  | R |
|---|---|
| plywood | 0.59 |
| urethane | 25 |
|  | 25.6 |

$U = 1/\sum R = 0.039$, where $\sum R$ means "the sum of all $R$s" in the path of heat flow.

Every square foot of ceiling transmits 0.039 Btuh/°F into the cold storage area. If the cold surface of the ceiling is at a temperature of $-25$°F and the exposed surface of the plywood at $+65°F$, then the temperature difference across the ceiling is $90°F$. The heat conducted into the cold storage area per square foot of ceiling is $90 \times 0.039$, or 3.5 Btuh.

All these thermal factors, $K$, $C$, $R$, and $U$, are per square foot and per $°F$. $K$ factor is also per inch thickness.

## TABLE OF THERMAL CONDUCTIVITIES

|  | $K$ (Btuh/ft²-°F-in.) |
|---|---|
| standard construction materials | |
| asbestos millboard | 5 |
| asphalt | 8 |
| common brick | 5 |
| concrete | 12 |
| glass | 3.6–7.2 |
| lightweight concrete | 0.70–1.15 |
| perlite plaster | 1.42 |
| soils | 4–12 |
| metals | |
| pure copper | 2700 |
| pure aluminum | 1400 |
| mild steel | 325 |
| stainless steel | 105 |
| insulating materials | |
| cellular glass | 0.35–0.41 |
| fiberglass board | 0.21–0.26 |
| foamed polystyrene | 0.24 |
| foamed polyurethane | 0.12–0.17 |
| mineral wool blankets | 0.27 |

## 2.3 surface coefficients of heat transfer

If heat is to be transferred from room air to a wall, or from a building wall to outside air, there must be a difference in temperature between wall and air, as indicated by Fig. 2.3. There is a narrow boundary layer of air with a small temperature drop across it. Heat transfer through this boundary layer of insulating air is a part of the heat path, and the heat flow through it (equal to the heat flow through the wall) is

$$Q = hA(t_a - t_w)$$

where $A$ = wall area

$t_a$ = ambient air temperature

$t_w$ = wall temperature

$h$ = air film coefficient, Btuh/ft² — °F

This surface coefficient $h$ is

1.65 for still air

6.0 for air with a 15-mph wind

These air film resistances on each side of a wall must be included in the overall coefficient of heat transmission $U$, thus:

$$U = \frac{1}{\sum 1/h + \sum L/K} = \frac{1}{\sum 1/h + \sum R}$$

$$Q = UA\,\Delta t$$

**Example.** A cold storage ceiling 800 ft² in area is built of $\frac{3}{4}$-in plywood ($K = 1.70$) and 4 in. of foamed polyurethane ($K = 0.16$). Film coefficient $h$ for both sides of the ceiling is 1.65. Determine the heat flow into the cold storage room through the ceiling, if the air temperature above the ceiling is 75°F and the cold storage area is maintained at −30°F.

$$U = \frac{1}{\sum 1/h + \sum L/K}$$

$$= \frac{1}{\dfrac{1}{1.65} + \dfrac{1}{1.65} + \dfrac{\frac{3}{4}}{1.70} + \dfrac{4}{0.16}}$$

$$= 0.0375$$

Heat flow through the ceiling $= UA\,\Delta t$

$$= 0.0375 \times 800 \times 105$$

$$= 3150 \text{ Btuh}$$

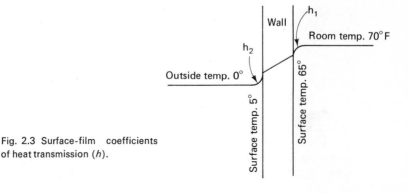

Fig. 2.3 Surface-film coefficients of heat transmission ($h$).

## 2.4 *vapor insulation*

Many of the activities carried on within buildings greatly increase the water-vapor content of the air within the building. Such activities include washing of floors, laundry work, cooking, washing of dishes, and the use of sinks and bathtubs. Since the air within a building is warm, it can hold a considerably greater amount of water vapor than colder air outside the building. The water-vapor pressure is therefore higher inside the warm building than in the colder outside air, and vapor therefore will tend to migrate through the material of walls and roof from inside to outside.

The most severe vapor migration problems are in cold climates, where temperature differences between inside and outside may be 100°F or even more. Temperatures of −40°F have been recorded at Devil's Lake, North Dakota, which is 110°F below standard room temperature of 70°F. Such temperature differences result in vapor-pressure differences of about 0.20–0.25 psi. This pressure difference will cause water vapor to penetrate the walls and roof of a building until it reaches some location within the structure which is at the dew-point temperature or temperature of condensation. Here water can collect. The condensed water can freeze and destroy roofing and insulating materials.

To prevent such condensation within the building structure, a vapor-barrier material is used. This is a layer of material that is impermeable or relatively impermeable to water vapor. The vapor barrier must be located on the warm side or inner side of the structure, that is, on the hot side of the dew-point position (Fig. 2.4).

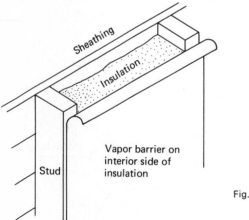

Fig. 2.4 Vapor barrier.

The effectiveness of a vapor barrier is rated in *perms*. An effective vapor barrier should have a vapor permeance rating that is not greater than 0.2 perm. A rating of 1 *perm* means that 1 ft² of the barrier is penetrated by 1 grain of water vapor per hour under a pressure differential of 1 in. of mercury (in. Hg). One inch of mercury equals 0.491, or virtually 0.5 psi. One grain is 1/7000 lb.

**Example.** The previous example of a cold storage ceiling will be analyzed for vapor migration. A vapor-pressure difference of 0.2 psi will be assumed, which is 0.4 in. Hg. The vapor barrier is rated at 0.2 perm. The ceiling area is 800 ft². The amount of moisture passing through the vapor barrier into the ceiling material will be estimated.

Note that accuracy is unimportant in this calculation; what is wanted is the approximate amount of vapor in order to decide whether it is a serious enough amount to cause damage.

The barrier admits 0.2 grain per square foot per hour per inch mercury pressure difference. For 0.4 in. Hg pressure, 0.08 grain per square foot-hour enters the ceiling insulation. Over the whole ceiling area of 800 ft², this is 64 grains per hour.

$$7000 \text{ grains} = 1 \text{ lb}$$

$$64 \text{ grains} = 0.01 \text{ lb (closely)}$$

This does not appear to be a significant amount of water (actually ice, since the temperature of the urethane insulation is presumably below freezing temperature). But the cold storage area must be maintained at $-30°F$ over the 24 hr of each day if food is stored in it. In a full day, nearly $\frac{1}{4}$ lb of ice may collect in the ceiling. Over a period of 1 yr, the deposit of ice could possibly break up the insulation or separate it from the plywood backing. At any rate, the standard perm rating of 0.2 perm seems chancy for this installation; what is really required is a virtually *perfect* vapor barrier. This statement can be supported by experience. This ceiling, cited in several examples in this chapter, is an actual installation. One of the electrical outlets in the ceiling admitted vapor to the cold storage area; in the first week of operation a chunk of ice 7 in. long grew from the ceiling at the edge of the electrical fixture. A cold storage room held at $-30°F$ of course presents an unusually severe vapor-barrier requirement.

A vapor barrier must be a continuous surface without holes or tears to admit vapor. When sheets of polyethylene or similar vapor-barrier material are used, they should be overlapped at least 6 in. or on studs should be lapped a full stud spacing. On subfloors the vapor barrier should be carried a few inches up the sidewalls. The barrier must be fitted carefully around wall openings such as electrical boxes. Similarly, paint coats used as vapor barriers must have an unbroken surface. Not all paints are adequate as vapor barriers, but aluminum paints and paints based on asphalt or oil bases are often suitable.

Moisture can enter a building through a concrete slab on grade. Such moisture may harm or detach floor tile laid on the concrete. To prevent such moisture penetration, a moisture barrier is laid on the earth under the concrete. Any vapor barrier will also serve as a moisture barrier; polyethylene film 4 or 6 mils thick (1 mil = 0.001 in.) is usually employed. A layer of sand is used as a cushion for the film, which must not be penetrated by concrete reinforcing or the concrete pouring operation.

The use of moisture-penetrating roofing materials can result in expensive damage. The moisture may travel through joints in the roofing and leak back into the building at points far removed from its origin. Trapped moisture in a roof may reevaporate during the summer under a hot roof and raise blisters or delaminate the roofing materials.

A built-up roof usually attains a barrier rating of 0.2 perm. Polyethylene and vinyl film, rubber sheets, waxed kraft paper, bitumen-coated paper, and aluminum foil are effective vapor barriers. Steel decks with caulked joints are equally effective.

There are cases where a vapor barrier creates worse problems than it solves. Obviously a vapor barrier will also serve to seal in moisture, and this may be a source of building problems. Where possible, therefore, the vapor barrier should be omitted. This is possible in milder climates and in unheated buildings.

|  | Perm Ratings |
| --- | --- |
| $\frac{3}{4}$ in. white pine | 1.4 |
| $\frac{1}{4}$ in. fir plywood | 4.5 (varies) |
| $\frac{1}{4}$ in. fir plywood, two paint coats | 2 |
| tempered Masonite, 0.13 in. | 4.8 |
| 15-lb asphalt roofing felt | 6.6 |
| 4 in. common brick | 0.8 |
| $\frac{3}{4}$ in. plaster | 13.3 |
| Elastron rubber roof coating, 15 mil | 0.03 |
| 3.8 in. gypsum board | 34.5 |
| kraft paper | 82 |
| aluminum foil, 1 mil | 0.0 |
| polyethylene, 2 mil | 0.16 |
| 1 in. foamed polyurethane | 3.5 |

## PERTINENT ASTM SPECIFICATIONS

E84-68 (Part 14) Surface Burning Characteristics of Building Materials
E96-66 (Part 14) Water Vapor Transmission of Materials in Sheet Form
E108-58 (Part 14) Fire Tests of Roof Coverings
E119-67 (Part 14) Fire Tests of Building Construction and Materials
E286-65T (Part 14) Surface Flammability of Building Materials
E176-66 (Part 14) Terms Relating to Fire Tests of Building Materials

1 Does the thermal conductivity of a material such as wood or steel have an influence on its fire resistance?

2 Stainless steels have about the same yield stress and modulus of elasticity as standard construction steels. If stainless steel were used instead of construction steels in buildings, would it likely have a better or worse fire rating than standard steels?

3 Determine the heat flow in Btuh through a 4-in. thickness of mineral wool blanket 3 × 12 ft in area, with its hot face at 65°F and its cold face at 0°F.

4 Consult the table of thermal conductivities in Section 2.2. Stainless steel cookware is actually a sandwich of mild steel with surfaces of stainless steel. Would the user of such cookware be better served if the material were solid stainless steel?

5 A 1000-gal water storage tank is insulated with polyurethane foam $\frac{1}{2}$ in. thick, with a $K$ factor of 0.16. Total surface area is 200 ft². The inside surface of the insulation is at a temperature of 35°F and the outside surface at −20°F. If 3412 Btuh = 1 kilowatt-hour (kwhr), how many kilowatts of electric heat are required to maintain these conditions?

6 A solid brick wall is 12 in. thick, with an area of 8 × 20 ft. If the temperature difference across the wall is 80°F, what is the heat flow through the wall?

7 Find the conductance of the following materials: (a) 3 in. of polyurethane foam with a $K$ factor of 0.11 Btuh; (b) $\frac{3}{4}$-in. thickness of plywood, $K = 1.1$; (c) 2 in. of foamed polystyrene; (d) 4 in. of mild steel.

8 Find the $R$ factor for the materials of Question 7.

9 Find the $U$ factor for a wall construction of $\frac{3}{4}$-in. plywood, $R = 0.60$, and 3 in. of mineral wool blanket. Do not include surface coefficients.

10 Find the $U$ factor for $\frac{1}{8}$ in. of asbestos millboard backed with $1\frac{1}{2}$ in. of foamed polystyrene.

11 What is the $U$ factor for a composite construction of 3 in. of concrete, 4 in. of foamed polystyrene, 3 more in. of concrete, and $\frac{1}{2}$ in. of asphalt?

12 Polyethylene film is used as temporary protection against the weather during construction. What is the heat loss per 100 ft² of polyethylene film for a temperature difference between inside air and film of 30°F and 30°F between film and outside air temperature? Use an inside surface coefficient of 1.65 and an outside coefficient of 6.0.

13 In Question 12, why does it not matter for heat flow how thick the film is?

14 For the insulated tank of Question 5, the outside air film coefficient is 6.0. What is the temperature of the outside air?

15 For the brick wall of Question 6, the inside air film coefficient is 1.65. What is the temperature difference between the inside surface of the brick wall and the air in the room enclosed by the brick wall?

16 Explain the need for a vapor barrier in construction.

**17** What harm might a vapor barrier do in building construction?

**18** Define a perm.

**19** Why must a vapor barrier be located on the hot side of a wall?

**20** How many grains in a pound?

**21** This book suggests that an effective vapor barrier should have a rating of not more than 0.2 perm. Other books suggest than 1 perm is adequate. What do you think?

**22** Suggest some methods of installation that could make a vapor barrier almost worthless.

**23** A vapor barrier is rated at 1 perm, and the vapor pressure difference across a 100-ft$^2$ wall is 0.5 in. Hg. In a day of 24 hr, how many pounds of water vapor could penetrate the vapor barrier under these conditions?

**24** For fire protection of the steel columns of the United States Steel Building in Pittsburgh, the columns are filled with water, and other fireproofing is not used.

(a) A steel structure will collapse long before the temperature of the steel reaches 1000°F. How many Btus of heat will be absorbed by a lineal foot of steel column if its temperature in a fire is raised by 1000°F? Assume that the column weighs 450 lb/lineal foot and the specific heat of the steel is 0.1 Btu/lb-°F.

(b) The same column encloses 9 ft$^3$ of water per lineal foot of column. What will be the temperature rise of this water if it must absorb the heat calculated in part (a) of this question? (1 ft$^3$ of water = 62.4 lb; specific heat of water = 1.0 Btu/lb-°F.)

(c) How will you prevent the water in such a column from corroding the column? Heating boilers are made of steel alloys similar to construction steels. Find out how boiler corrosion is prevented.

(d) Besides leakage of the water from the column, there is the problem of expansion of the water in the column as it is heated. Does this present problems?

Acoustics, the science of sound, is a complex science and a major study in itself, Nevertheless, some general principles regarding this subject are easily understood. Acoustical treatments have become increasingly important in the design of buildings. A building is an enclosed environment, and sound is a part of that environment.

In discussing the acoustic properties of materials, we are primarily interested in the reflection, transmission, and absorption of sound by construction materials, that is, reflectivity, transmissivity, and absorptivity. Two quite different acoustic considerations are involved in the control of sound within buildings:

1. The absorption and reflection of sound by the material, that is, the effect of materials on the acoustical conditions within a room.
2. The transmission of sound by materials, that is, the transmission of sound from one room to an adjoining room.

# ACOUSTICS
# AND CONSTRUCTION
# MATERIALS

# 3

When a sound wave impinges on a surface, some of its energy must be reflected, some absorbed, and some transmitted through the material. All the incident acoustic energy is accounted for in these three effects (see Fig. 3.1)

$$R + A + T = 1$$

where $R$ = fraction of sound energy reflected (*reflectivity*)

$A$ = energy fraction absorbed and dissipated within the absorbing material (*absorptivity*)

$T$ = energy fraction transmitted through the material to the adjoining room

These acoustical properties of materials depend upon other more fundamental material properties, including modulus of elasticity, density, mass, hardness, porosity, elasticity, and plasticity. Since concrete floors are highly reflective to sound energy, while carpets are much less so, it is known intuitively that hard materials are highly reflective to sound, while soft and porous materials are less reflective.

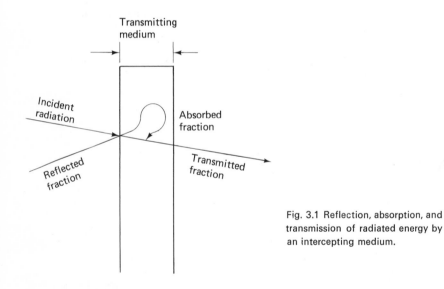

Fig. 3.1 Reflection, absorption, and transmission of radiated energy by an intercepting medium.

These acoustical properties are also dependent on the frequency (pitch) of the sound wave. The *frequency*, or *pitch*, is the number of wavelengths of vibration that pass a fixed point in 1 sec. Low frequencies are in the range below about 100 hertz (Hz; 1 Hz-100 cycles per second). This is just below the humming sound of a 60-Hz electric power transformer, which emits 120-Hz sound. High frequencies are those ranging up to about 12,000 Hz, which is the approximate limit of hearing for most persons, although it is assumed that the audible range goes to 20,000-Hz sound. Beyond the audible range lie the ultrasonic frequencies, which are not of concern in construction.

Higher frequencies of sound are usually characterized by smaller amounts of energy and are more easily attenuated or absorbed. As one moves farther and farther away from a dance orchestra, a distance is reached at which the higher tones of the music can no longer be heard, although the low-frequency rhythmic beat of the orchestra can still be sensed. Absorption of sound by acoustic tile or other types of sound-absorbing material is therefore always more effective for higher frequencies. The increase in sound absorption with frequency can be illustrated by the case of an ordinary plaster wall:

### ACOUSTIC ABSORPTION COEFFICIENTS FOR 1/2 IN. PLASTER

| Low Frequency (125 cycles) | Medium Frequency (500 cycles) | High Frequency (4000 cycles) |
|---|---|---|
| 0.013 | 0.025 | 0.045 |

Thus 1.3 percent of the incident sound energy is absorbed and transmitted by the plaster at low frequencies, 4.5 percent at high frequencies.

## 3.2 the decibel

The intensity of a sound wave is the amount of energy in the sound, or, in everyday language, the loudness. Intensity is measured in *decibels* (*db*), or tenths of bels. An intensity of 0 db is the least sound that is discernible to the human ear, and corresponds to an energy of $10^{-16}$ watt per cm². The higher the decibel level the more powerful is the sound. Long exposure to decibel levels of 90 or more causes temporary or permanent deafness. Figure 3.2 displays the decibel levels of some sounds and environments.

The bel represents a ratio of 10:1:

$$\text{bels} = \log_{10} \frac{\text{quantity measured}}{\text{reference quantity}}$$

and since decibels are tenths of bels,

$$\text{decibels} = 10 \log_{10} \frac{\text{quantity measured}}{\text{reference quantity}}$$

The decibel is therefore a logarithmic unit. The reason for such a choice of unit is simply that the human ear responds in an approximately logarithmic manner to changes in intensity of sound.

| Decibels | Sound | Effect |
|---|---|---|
| 120 | Riveting Pavement breaker Boiler factory | Deafening |
| 100 | | |
| | Loud street noise Noisy factory | Very loud |
| 80 | | |
| | Average street noise Noisy office Average factory | Loud |
| 60 | | |
| | Noisy home Average office Conversation Quiet radio | Moderate |
| 40 | | |
| | Quiet home Private office Quiet conversation | Faint |
| 20 10 0 | Rustle of leaves Whisper Threshold of audibility | Very faint |

Fig. 3.2 Acoustic sound levels. (Modified from Celotex Corp. chart.)

As an example, consider an intensity of sound to be doubled:

$$\log_{10} \tfrac{2}{1} = 0.3.$$

This is a 3-db increase in sound intensity, which is about the least increase that is perceptible to the human ear. The following table converts absolute intensity levels to decibels:

| Energy Intensity (watt/cm$^2$) | Decibels |
|---|---|
| $10^{-16}$ | 0 |
| $10^{-15}$ | 10 |
| $10^{-14}$ | 20 |
| $10^{-13}$ | 30 |
| $10^{-12}$ | 40 |
| $10^{-11}$ | 50 |
| $10^{-10}$ | 60 |
| $10^{-9}$ | 80 |
| $10^{-6}$ | 100 |
| $10^{-4}$ | 120 |

## 3.3 sound generation
## in buildings

Two general types of sound generation must be provided for in building construction. One type comprises those sources which are airborne, examples being the human voice and loudspeakers. The other type includes those sources which act directly on the structure of buildings and which are transmitted through the structure. Some examples of the second type are footsteps, banging doors, vibrating machinery, dropping of objects on floors, and repositioning of furniture. This second type of sound generation is often termed "impact" sound generation, since most structure-borne noises result from impacts.

## 3.4 airborne sound

When the pressure fluctuations of airborne sound act upon a wall, they force the wall into vibration at the frequency of the sound wave (see Fig. 3.3). The wall therefore acts as a loudspeaker to generate sound waves at the rear of

Fig. 3.3 Impact and airborne sound.

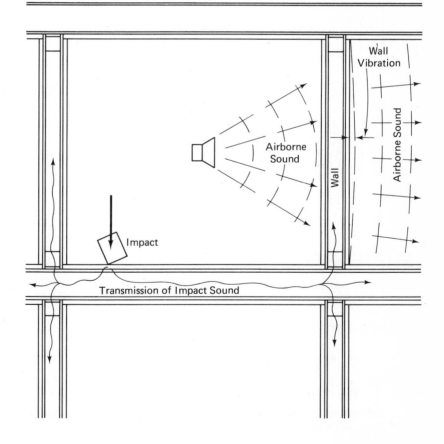

the wall, that is, in the adjoining room. Airborne sound in the adjoining room will be reduced by designing a wall that is difficult to set in motion. The ratio of the sound energy falling upon one side of a wall to the sound energy transmitted by the wall, expressed as decibels, is called the *transmission loss*. The energy not transmitted is either absorbed by the wall or reflected A higher decibel number therefore indicates better sound insulation. A heavy wall is more difficult to set into vibration and provides less transmission than a lightweight wall, since force = mass × acceleration, and the greater the mass, the less the acceleration caused by the acoustic force. Higher frequencies are less effective in setting up induced vibrations in a structure. Therefore, the larger the product of wall mass times frequency, the more effective is the sound insulation. This rule is known as the *Mass Law*. Partitions of sandwich construction or lightweight semisandwich construction, such as wallboard on studs, provide poor sound insulation, although they offer very effective thermal insulation. Transmission of sound can occur through the studs of

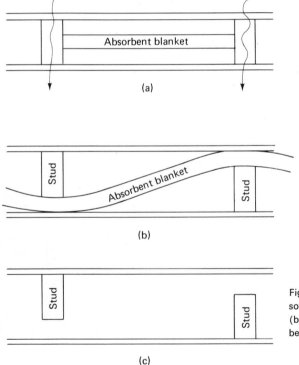

Fig. 3.4 Sound transmission: (a) sound leakage path through studs; (b) improved installation of absorbent material; (c) staggered studs.

the wall, and for better sound insulation staggered studs are used, as in Fig. 3.4. The effectiveness of soundproofing requires attention also to a number of subsidiary details, such as flanking transmission through adjacent structural elements and sound leaks of the type shown in Fig. 3.5.

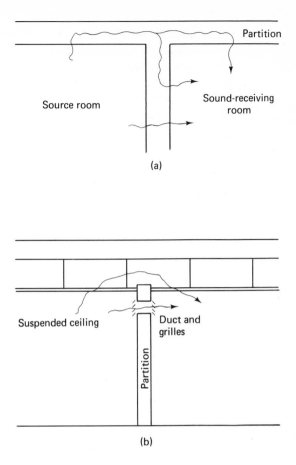

Fig. 3.5 Sound leaks: (a) flanking transmission through adjacent structure; (b) sound leaks through air spaces.

## 3.5 impact sound

Impact sound may be carried considerable distances through a building constructed of rigid materials such as concrete or steel. The sound must be radiated by walls, floors, or other efficient sound radiators before it can be heard. The effects of impact sound are reduced by the use of resilient floor tile and carpets and the use of flexible mountings on vibrating machinery such as ventilating fans. Finished floors may be laid over absorbent materials such as fiberglass; suspended ceilings may be hung on resilient hangers.

## 3.6 sound absorption

Various types of surface materials, such as acoustic tile, are applied to walls and ceilings. Such materials have little effect on the sound transmission of the wall to which they are attached. They are not good reflectors of sound. Their use is to absorb sound from the air of the room.

Sound-absorbing materials convert acoustic energy to frictional heat. Fabrics, carpets, and upholstered furniture are porous. The air within such surfaces is pumped back and forth by the acoustic waves, with the result that some of the sound energy is lost as frictional heat moves through the restricted passages of such material. A similar effect is produced by the holes in acoustic tile. Such materials are more effective at high frequencies than at low frequencies.

The ratio of acoustic energy absorbed (converted into heat) to the energy falling on the surface is called the *sound absorption coefficient*. Since this ratio will increase with frequency, the sound absorption coefficients measured at 250, 500, 1000, and 2000 cycles are averaged, this average being called the *noise reduction coefficient*. Note that these coefficients measure the ratio of acoustic energy *not* reflected to the incident energy. An open window is a perfect absorber, with a coefficient of 1.0.

### NOISE REDUCTION COEFFICIENTS FOR POROUS MATERIALS

| | NRC |
|---|---|
| mineral wool blankets, $\frac{1}{2}$–4 in. thick | 0.45–0.95 |
| acoustic plaster, $\frac{3}{8}$–$\frac{3}{4}$ in thick | 0.25–0.40 |
| carpets | 0.30–0.60 |
| draperies | 0.10–0.60 |

## 3.7 acoustically absorbent materials

The most familiar of the acoustic materials is acoustic tile, most of which are intended for ceilings. These are soft and fibrous materials, of wood, asbestos, or other fibers bonded into tile of various thicknesses. The edges of the tile may be squared, or beveled, or grooved for tongue-and-groove joints. A size of 12 × 12 in. is popular, although other sizes are also in use, in multiples of 12 or 16 in.

Such ceiling tile may be attached by a variety of methods. If attached to a solid ceiling or to furring strips, the tile may be nailed or clipped. Mastic may also be used to bond the tile, especially if the tile is to be attached to concrete or gypsum board. Four dabs of mastic about 2 in. in diameter per 12-in. tile are applied to the back of the tile. Another method of installation is to hang the tile in a suspended ceiling frame, as shown in Fig. 19.53.

The acoustic absorption of these materials is improved by the pattern of holes drilled or punched in the surface of the tile (see Fig. 3.6). Such tiles are painted at the factory and do not require painting on site. On-site painting causes a slight loss in acoustic absorption, or a more serious loss if the perforations are closed with paint. The noise reduction coefficient of such materials is of the order of 0.70.

Fig. 3.6 Acoustical treatment of a wall for high-frequency engine noise: perforated hardboard below, with acoustical tile above.

A type of acoustic surface somewhat resembling tile is a larger assembled sandwich unit with a perforated hardboard face and a fiberglass or rockwool backing. Air with its associated sound is admitted to the porous backing material through the perforated front panel, and acoustic energy is dissipated by frictional effects.

Sprayed or troweled acoustic surfaces may be an acoustic plaster with a foaming agent or a mineral fiber mixed with adhesive. These materials are acoustically effective because of their porosity, giving absorption coefficients in the range 0.70–0.80. Acoustic plasters may use a vermiculite or perlite aggregate with gypsum and a foaming agent to provide porosity. These are finish plaster coats, using two coats to provide a total thickness of $\frac{1}{2}$ in. of acoustic plaster. Mixing is done on site, and the manufacturer's recommendations should be carefully followed if an effective acoustic plaster is to be produced. Machine spraying is preferred if plastering must be done on irregular surfaces.

Mineral fibers are also sprayed. These contain an inorganic binder, and the surface to be sprayed requires a priming adhesive.

## QUESTIONS

1  A material has an acoustic absorption coefficient of 0.65. What fraction of the sound energy is either transmitted or absorbed by the material? What fraction is reflected back to the room?

2  The foam cells of foamed polystyrene (styrofoam) do not interconnect with each other. Will such materials make effective sound absorbers?

3  Suggest several examples of airborne and structure-borne sound.

4  Sound is not really transmitted through a wall; it is reradiated by the wall. Explain.

5  What is the meaning of the Mass Law for sound?

# THE
# MATERIALS

*Part*

The weight of all structures must be carried either on soil or on rock. Clay soils and silts are composed of eroded particles of rock, while large pieces of rock are called stone. Both soils and rock are also used as raw materials for building products. Rock is quarried to make building stone or fragmented to produce loose fill and aggregate for concrete. Suitable clays are used in the manufacture of cement, brick, and other products.

Concrete is a manufactured rock, and rock and concrete are similar in some of their physical properties. But since concrete is manufactured, a high degree of quality control is possible, while the properties of rock must be accepted as found. The E values of rocks are greater than those determined for concrete. The tensile strength of both concrete and rock specimens is about 10 percent of the compressive strength. However, it is not generally possible to produce concretes that are as strong in tension or compression as rock. Strength tests of rock specimens are not necessarily representative of the properties of the rock mass as it lies in the earth's crust, since the extracted rock specimen is relieved of the compressive pressures and the continuity of the rock deposit in which it lay.

# ROCK
# AND
# SOIL

# 4

## 4.1 geological classes
## of rock and stone

Rock is classified by its geological origin into three types: igneous, sedimentary, and metamorphic.

*Igneous rock* is formed by the solidification of molten material, usually at some depth in the earth's crust, or at the earth's surface by volcanic or other action. Granites are examples of igneous rock and are chiefly composed of quartz, feldspar, orthoclase, and smaller amounts of mica. Igneous rocks are subdivided according to their chemistry as acid, intermediate, and basic (i.e., alkali), these subdivisions being determined by the amount of silica present in the rock. Granite has a high content of quartz, which is a silica, and so is acid. The high quartz content makes granite a hard rock.

*Sedimentary rock* is formed by the deposition, usually under water, of mineral matter chiefly produced by the destruction of preexisting igneous rocks. Sandstone, for example, is a sedimentary rock of cemented quartz particles, the cementing material being usually calcite. Shale is deposited from clay. Limestone is a calcium carbonate and often contains fossils. Dolomite is a mixture of calcium and magnesium carbonate. All these sedimentary rock types tend to be weaker than igneous rock and are usually jointed and stratified.

*Metamorphic rock* is either igneous or sedimentary rock that has been changed from its original structure by extreme heat or pressure at some period in its geological history. Marble, quartzite, and slate are examples of metamorphic formations. Slate is formed from clays and shales. Marble is a recrystallized limestone or dolomite.

## 4.2 hardness

Rocks are basically aggregates of mineral particles. Hardness is an indication of strength in rock, although the strength of a rock is in part dependent on particle size, smaller grain size resulting in a stronger rock. The strength of a rock is also influenced by structural imperfections such as porosity, cracks, inclusions, and even weak particles.

The usual hardness scale for rock and stone is *Moh's scale* of 10 minerals, each of which will scratch the next lower grade:

### MOH'S SCALE OF HARDNESS

| | | | |
|---|---|---|---|
| 10 | diamond | 5 | apatite |
| 9 | corundum | 4 | fluorspar |
| 8 | topaz | 3 | calcite |
| 7 | quartz | 2 | gypsum |
| 6 | feldspar | 1 | talc |

The tungsten carbide inserts in rock drills have a hardness between 8 and 9. Tool steel is harder than quartz but will not scratch topaz, number 8. Glass has a hardness of 5 on Moh's scale.

Moh's scale is based on certain minerals, whereas rock and stone are agregates of mineral particles often bonded with a different mineral. Deposits of silica sand, for example, are composed of particles of hard silica bonded with soft kaolin clay. Hence the hardness of a mineral may be known, but the hardness of a rock is ambiguous. Despite the hardness of the grains, it is possible to cut silica sand with a knife. Probably the hardness of the grains should be termed the *hardness* of the rock, and the strength of the bond should be termed the *toughness*. Toughness of rock and stone can be roughly measured by the ease with which the material can be broken with a hammer.

In the case of commercially used stone, the hardness and the strength are proportional to the silica content. The cost of stone is also roughly proportional to the silica content, because the difficulty of drilling, cutting, polishing, and working the stone increases with silica content. Thus limestone, which usually contains little silica, is easy to drill or process, and is called *soft rock* in the mining industry. Quartz and quartzite are almost pure silica and are thus very hard and abrasive, or *hard rocks*. Granites are siliceous and are therefore expensive. The durability of a stone for building purposes may be thought to be roughly proportional to the silica content, but there are other factors to consider, such as porosity.

The author once had to excavate in a hard quartzite formation in a northern latitude. The rock was hard to drill, but it fragmented well when blasted. It had an unusual gray-blue color which could have attracted an imaginative architect. The natives called this rock "growing rock" because it seemed to grow in size. The rock was heavily fissured, and frost action would break it into large fragments. Frost movements would cause these fragments to lever themselves up until, over a period of years, a pile of rock would reach 6 or 8 ft above the level at which it lay originally. Such a rock would be a risky building stone, despite its attractive color.

## 4.3 porosity

All rock is porous to some degree. *Porosity*, the percentage of a rock that is voids, has a powerful influence on the physical characteristics of a rock or stone, especially strength or weatherability. The pore spaces are usually continuous and in the form of irregular cracks of microscopic size that separate the mineral grains. The degree of porosity depends on the method of formation of the rock. A rock formation that cooled slowly from the liquid condition will be relatively nonporous; a rapidly cooled lava that released gases during cooling will be quite porous. Sedimentary rocks are more porous than igneous or metamorphic rocks. In the table that follows, shale, limestone, and dolomite are all sedimentary; their high porosity will be noted. The higher porosity of shale is explained by its lack of compaction.

| | Specific Gravity (water = 1.0) | Porosity (%) |
|---|---|---|
| granite (igneous) | 2.6–2.7 | 0.5–1.5 |
| sandstone (sedimentary) | 2.0–2.6 | 5–25 |
| shale | 2.0–2.4 | 10–30 |
| limestone | 2.2–2.6 | 5–20 |
| dolomite | 2.5–2.6 | 1–5 |
| marble (metamorphic) | 2.6–2.7 | 0.5–2 |
| quartzite | 2.65 | 0.1–0.5 |
| slate | 2.9–2.7 | 0.1–0.5 |

## 4.4  building stone

Building stone is used in the form of (1) rubble stone; (2) dimension stone or cut stone, cut to specified size; (3) flagstone or thin slabs; and (4) crushed rock for concrete aggregate or other purposes.

Stone is associated with timeless durability, witnessed by such monumental structures as the pyramids of Egypt and the medieval cathedrals of Europe. Although rock is abundant in most geographical regions, relatively few deposits are acceptable for construction purposes. Some deposits are unsuited for reasons of inaccessibility, poor weatherability of the rock, low strength, low hardness, or unattractive appearance. Hardness is clearly significant for floors, walks, and stair-tread installations, but it also has a powerful influence on the workability of the stone and therefore on its cost. Color and texture (fineness of grain) influence the relative beauty of the stone.

*Granite* is hard, strong, and durable and can be given a fine polish. It is found in a range of colors, including reds and pinks, grays, and browns. Quartzite sometimes resembles granite but is usually coarser in appearance.

*Limestone* and *dolomite* are strong sedimentary rocks, light gray in color. Travertine is a variant of limestone, having been formed by the evaporation of water from hot springs. It is used as an interior decorative stone, often in floors.

*Marbles* are well known for their range of colors and color patterns. Marble may be white, gray, or black, as well as pink, yellow, green, and other colors. Not all marbles are suitable for exterior applications.

*Serpentine* colors range from olive green to almost black. This stone can be given a very attractive polish, like marble, but is not usually suitable for exterior applications, although it is used for flagstones.

*Sandstone*, a porous sedimentary rock, may be buff, brown, gray, or red in color.

Fig. 4.1 Beauty in stone: A bell tower against the prairie sky, Grafton, North Dakota.

*Slate* is a stratified metamorphic rock easily broken into thin sheets called slates.

Besides polished and rubbed finishes, stone is given a variety of textured finishes: sawed finishes, machine tool finishes, and hand finishes.

## 4.5 construction in stone

The use of stone for bearing walls is now rare. Most stone is used as a facing material. Construction patterns include ashlar or rubblework, and some stone is also used as building trim. *Rubblework construction* uses stone that is not cut to size but has the exposed face of the pieces chipped. Rubble construction may be random or coursed, as shown in Fig. 4.2. Random work produces

Fig. 4.2 Random and coursed rubble construction.

neither horizontal nor vertical courses. In coursed work horizontal courses are used.

   *Ashlar construction* uses cut stone, in *coursed, irregular coursed,* or *broken ashlar.* These three methods are shown in Fig. 4.3.

Fig. 4.3 Coursed, irregular coursed, and broken ashlar.

   Stone may be used as trim for a variety of structural functions. In framing doors, stone may be used for the sill, the jams, or the lintel (see Fig. 4.4). Stone sills are made in two types, *slip sills* and *lug sills.* The lug sill extends underneath the jambs. The sill is always sloped to allow rainwater to run out of the doorway.

   *Quoins* are stones used at the corners of a building (Fig. 4.5). It is usual to use a half-bond in laying up quoins; this is shown in the figure.

Fig. 4.4 Stone sill, jambs, and lintel in a stone entrance of a style now obsolete.

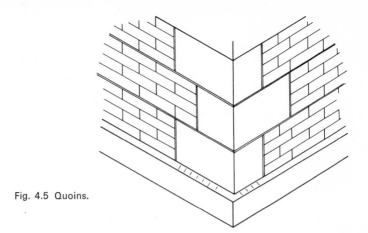

Fig. 4.5 Quoins.

*Cornice stones* provide the effect of an eave. A coping stone, shown in Fig. 4.6, is placed at the top of a wall or parapet to prevent water from entering the wall structure. The coping stone must be flashed with sheet metal and its upper surface must be sloped to shed water (Fig. 4.6).

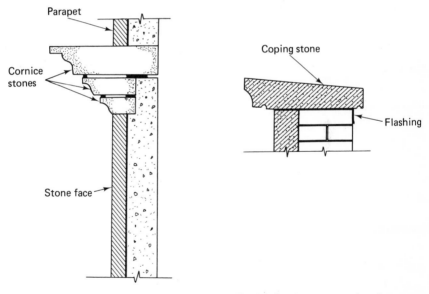

Fig. 4.6 Cornice stones and coping stones.

*Belt courses* of stone are special courses built into a wall to serve some purpose not served by the other courses of stone or brick (Fig. 4.7). Such a course may serve to provide a contrast to a large wall area of brick, as in the figure. A belt course may also be used at a change in wall thickness to provide a pleasing transition where the wall is stepped back.

Fig. 4.7 Belt courses and other features in stone for a large
armory.

When stone is used as a surface veneer, moisture must be prevented
from entering the rear surface of the stone by damp-proofing the backup
material.

Stone floors, patios, and walks are made of flagstones laid over a base
of gravel or concrete. The flagstones may be random (uncut shapes) or trim-
med (one or more sides saw-trimmed).

## 4.6 artificial stone and

### synthetic stone

*Artificial* or *cast stone* is a masonry product made of concrete. The exposed
face is given a decorative texture. Artificial stone that is split presents two
broken faces, the resulting rough surface giving an attractive appearance
when built into a wall of such units. Other textures may be smooth, polished,
or tooled, and colored or uncolored. The surface may also include granite,
quartz, marble, or other materials such as aggregate or chips.

Artificial stone facing used as veneer is produced in standard thicknesses
of $1\frac{1}{2}$, 2, and $2\frac{1}{4}$ in. Copings and stair treads of artificial stone are also produced.

*Synthetic stone* is made from minerals and a polyester plastic binder. Both synthetic marble and synthetic granite are produced. The mix for a synthetic granite approximates the following formulation:

| | |
|---|---|
| polyester resin | 20 lb |
| fine crushed granite | 75 lb |
| fine silica sand | 35 lb |
| | 130 lb |

The combination of crushed granite and silica sand produces the appearance of authentic fine-grained granite.

A black marble formulation would consist of the following:

| | |
|---|---|
| polyester resin | 20 lb |
| black pigment | 3 lb |
| fine silica sand | 110 lb |
| | 133 lb |

Synthetic granite contains about 40 per cent granite, but synthetic marble contains no marble. The veining effect of some marbles is easily imitated by a veining pigment mix of two or more colors.

These synthetic "stones" are produced in veneer slabs usually $\frac{3}{8}$ in. thick. They may be fastened to building walls with epoxy adhesive, or if fasteners are required they are soft enough to drill with ordinary drill bits. Slab size is usually 2 × 4 ft, since this size is a convenient one for a man to handle and is not so large that breakage becomes a serious cost problem. These stone veneers can be installed on a building for much less than the cost of natural stone. Granite, especially, is expensive, because of its hardness. In addition to these cost considerations, the synthetic stone mix is less heavy than natural stone and less expensive to ship and handle.

## 4.7 natural oxides

Rock and clay are complexes of six common oxides. These oxides are the basic materials for a wide range of industrial products in the civil and mechanical engineering areas. Other oxides may be present in smaller quantities as impurities. These six are the most important, however:

| | |
|---|---|
| silica | $SiO_2$ |
| alumina | $Al_2O_3$ |
| lime (calcia) | $CaO$ |
| magnesia | $MgO$ |
| iron oxide | $Fe_2O_3$ |
| rutile (titania) | $TiO_2$ |

*Clays*, for example, are complex aluminum silicates or complexes of silica and alumina.

*Silica* is found naturally as quartz, most often encountered as quartz rock, silica sand, sandstone, ore quartzite. This is the basic material for glass.

*Alumina* is the raw ore from which aluminum metal is extracted, usually the mineral bauxite.

Since silica is acid, and lime and magnesia are basic (alkaline), these materials readily react to produce calcium and magnesium silicates in geological formations, and calcium silicates in the manufacture of portland cement. *Lime* is produced from limestone, $CaCO_3$ or calcium carbonate, and *magnesia* from magnesite, $MgCO_3$. The mineral dolomite, often identical in appearance to limestone and also used as a building stone, is a solid solution of about half limestone and half magnesite. Lime is used in mortars, in the manufacture of portland cement and of sand-lime brick, in glass, sand-lime mortar, and in masonry paints.

Three simple chemical reactions are of great importance in the many uses of lime and mortars. *Quicklime* or *hot lime*, CaO, is made by dissociation of calcium carbonate at 1630°F:

$$CaCO_3 \longrightarrow CaO + CO_2$$

This process is called *lime burning*. *Hydrated lime* or *slaked lime* is produced by reacting hot lime with water:

$$CaO + H_2O \longrightarrow Ca(OH)_2$$

This slaking operation produces considerable quantities of heat. The slaked lime hardens in mortars by a slow reaction with carbon dioxide from the atmosphere:

$$Ca(OH)_2 + CO_2 \longrightarrow CaCO_3 + H_2O$$

thus reverting back to calcium carbonate.

Iron actually has two oxides. The red oxide of iron rust and most iron ores is $Fe_2O_3$, which is not magnetic. The black oxide found on hot-rolled steel products, called mill scale, is $Fe_3O_4$, which is magnetic. Iron oxides have limited industrial uses, one of the more important being a flux in the manufacture of portland cement. Iron oxides are the reason for the red color of clays in the southern states and the ocher color of many other clays. White clays such as kaolin are white because of the absence of iron oxide, and this characteristic makes such clays valuable. Iron oxide must not exceed 0.25 per cent in the silica sand used in glass manufacture, because iron oxide will color the glass. Hence for window glass or Coke bottles, the iron oxide content must be low, but the opposite is true for beer bottles or other brown bottles.

*Rutile* is one of the more common constituents of the earth's crust, although not so common as the ever-present silica and alumina. Rutile is a white oxide used to provide whiteness and opacity in paints and porcelain enamels. Rutile and lime are two of the more important ingredients in the flux coatings used on arc-welding rods.

## 4.8 soils

The foundations of all structures must be carried on the base materials, rock and soil.

The construction materials discussed in this book are usually materials produced to widely known and guaranteed standards. The steel mill that supplies reinforcing steel of 80,000-psi strength will certify that strength and will not ship steel which does not meet that strength. Construction plastics now have guaranteed flame spread ratings. Unfortunately, we do not have these guarantees where they matter most, in the rock and soil that must support the structure. The properties of the soil at the building site are totally unknown until investigated and are certain to change in the presence of the structure to be erected. If a building site is selected, the soil must be used as found. Often enough expensive building sites have been abandoned because of soil conditions. Abandonment of the site may be expensive, but it is cheaper than erecting an unstable building on dangerous soil. The most famous of foundation failures, the Leaning Tower of Pisa, is a tourist attraction. In the Arctic, small buildings have sunk out of sight into soils that could not support them.

Soil studies are a major consideration and a somewhat complex one. Such studies are a separate specialty and too large a subject for thorough discussion in a book on construction materials, especially since many of their significant properties are not matters of concern for other building materials.

The word "soil" as used in construction includes all inorganic and organic materials that bear the weight of structures, except solid bedrock. It is assumed that soil has at least some solid matter; in muck, muskeg, and permafrost there may not be much of it. Most soils are formed by the continual destruction and weathering of rock by the agencies of wind, water, and frost. Residual soils are those which have developed and remained in place over the rock from which they were formed. Such soil deposits are reasonably shallow. Foundation problems are rare in residual soils. Most soils, however, were transported either by water or wind. Some soils were transported by glacial action.

## 4.9 soil grading

Soils are divided into two main groups, coarse-grained and fine-grained. The coarse-grained soils are the *gravels* and *sands* and include all soils with particles visible to the naked eye. The grains of fine-grained soils are not visible. Almost always it is the fine-grained soils, the silts and clays, which give foundation difficulties.

Of the two types of fine-grained soils, the coarser fraction is called *silt*, and the finest fraction is *clay*. The unusual properties of clay, such as its plasticity when wet, result from the complex interaction between clay minerals and water. Silts are very fine and unaltered rock particles, while clays have been chemically altered.

Fig. 4.8  Soil classification systems: U.S. Department of Agriculture, Massachusetts Institute of Technology, and American Society for Engineering Education.

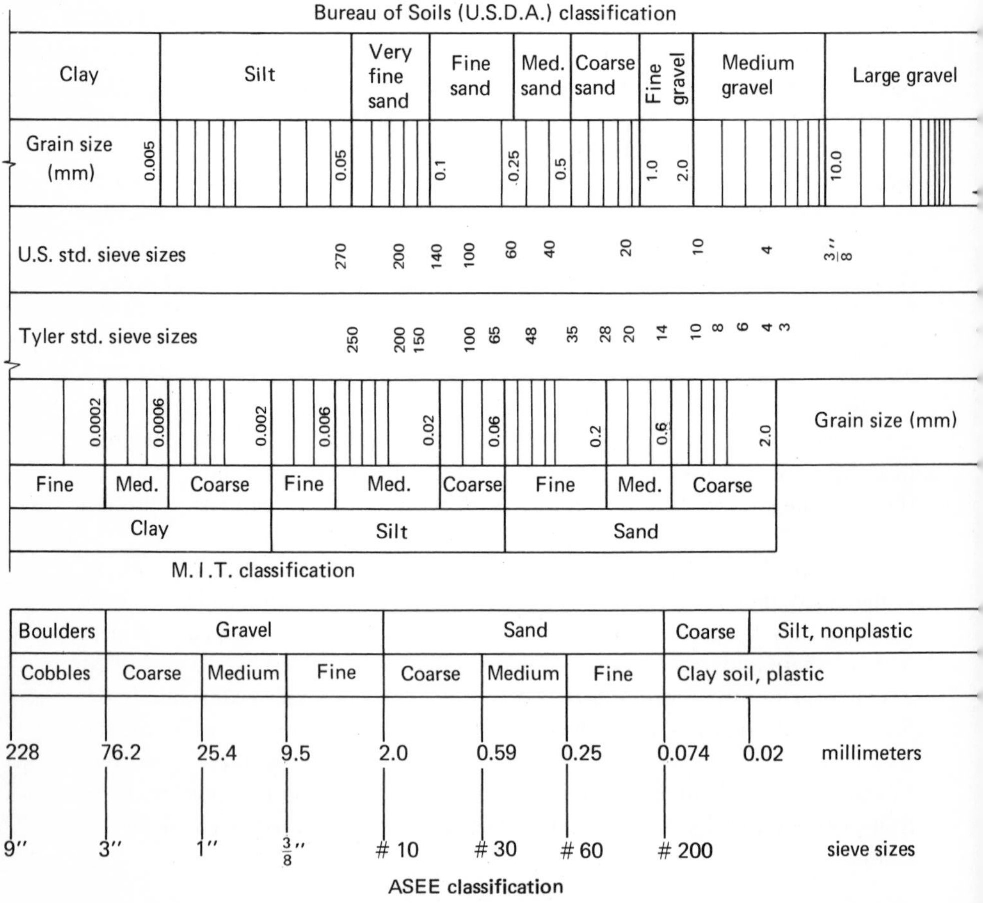

**Bureau of Soils (U.S.D.A.) classification**

| Clay | Silt | Very fine sand | Fine sand | Med. sand | Coarse sand | Fine gravel | Medium gravel | Large gravel |
|---|---|---|---|---|---|---|---|---|
| Grain size (mm) 0.005 | 0.05 | 0.1 | 0.25 | 0.5 | 1.0  2.0 | 10.0 | | |
| U.S. std. sieve sizes | 270 | 200 140 100 | 60 40 | 20 | 10 | 4 | 3/8 | |
| Tyler std. sieve sizes | 250 | 200 150 | 100 65 | 48 | 35 28 20 | 14 10 8 6 4 3 | | |

| | | | | | | | | Grain size (mm) |
|---|---|---|---|---|---|---|---|---|
| 0.0002 | 0.0006 | 0.002 | 0.006 | 0.02 | 0.06 | 0.2 | 0.6  2.0 | |
| Fine | Med. | Coarse | Fine | Med. | Coarse | Fine | Med. | Coarse |
| Clay | | | Silt | | | Sand | | |

M.I.T. classification

| Boulders | Gravel | | | Sand | | | Coarse | Silt, nonplastic |
|---|---|---|---|---|---|---|---|---|
| Cobbles | Coarse | Medium | Fine | Coarse | Medium | Fine | Clay soil, plastic | |
| 228 | 76.2 | 25.4 | 9.5 | 2.0 | 0.59 | 0.25 | 0.074 | 0.02  millimeters |
| 9" | 3" | 1" | 3/8" | # 10 | # 30 | # 60 | # 200 | sieve sizes |

ASEE classification

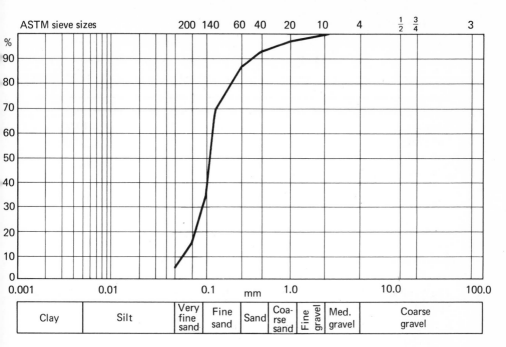

Fig. 4.9 Particle-size analysis for a fine sandy soil.

Figure 4.8 shows three classifications of soils by grain size. Note that there are slight differences between these classifications. Grain-size analysis is reported graphically as the cumulative percentages of various sieve sizes down to a 200-mesh sieve. Silt and clay particles are smaller than 200 mesh and must be sized by other methods, such as sedimentation. Figure 4.9 is such a particle-size analysis for a coarse-grained soil from the Ozark area of Arkansas. A nest of sieves for a particle-size analysis is shown in Fig. 5.6.

**Example.** A soil gives the following amounts retained on the sieves:

| Sieve Size | Weight Retained (g) |
|---|---|
| $\frac{3}{8}$ in. | 0 |
| No. 4 | 0 |
| No. 10 | 26.3 |
| No. 20 | 24.1 |
| No. 40 | 22.7 |
| No. 60 | 18.1 |
| No. 140 | 13.0 |
| No. 200 | 7.3 |
| passing 200 | 4.0 |
| | 115.5 |

The cumulative percentages are prepared as follows:

| Sieve Size | Weight Retained (g) | Weight Passing (g) | Cumulative Per Cent |
|---|---|---|---|
| $\frac{3}{8}$ in. | 0 | 115.5 | 100.0 |
| No. 4 | 0 | 115.5 | 100.0 |
| No. 10 | 26.3 | 89.2 | 77.3 |
| No. 20 | 24.1 | 65.1 | 56.5 |
| No. 40 | 22.7 | 42.4 | 35.8 |
| No. 60 | 18.1 | 24.3 | 21.0 |
| No. 140 | 13.0 | 11.3 | 9.8 |
| No. 200 | 7.3 | 4.0 | 3.5 |
| passing 200 | 4.0 | 0.0 | 0.0 |
| | 115.5 | | |

The results are plotted in Fig. 4.10. Note that the grain size scale is logarithmic instead of linear in order to cover the great range of grain sizes. Such a graph, linear vertically and logarithmic horizontally (or vice versa), is called *semi-logarithmic*.

Fig. 4.10  Soil grading example.

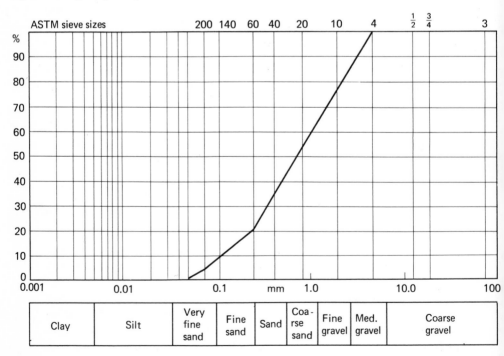

The difference between a sand, a silt, and a clay is determined by the following rough field tests:

| Fine Sand | Silt | Clay |
|---|---|---|
| visible particles | particles rarely visible | no visible particles |
| crumbles readily when dry | crumbles readily when dry | difficult to crumble when dry |
| feels gritty | feels rough | feels smooth |
| no plasticity | limited plasticity | distinctly plastic |

If a small piece of soil is broken after it has been dried, its breaking strength can be used for identification. If strong, the sample is almost certainly a clay. A silt tends to powder readily.

Another test is the shine test. A moist sample of soil is stroked hard with the flat side of a knife. If this produces a shiny surface, the sample is a clay. A silt gives a dull surface.

The limits of soil particle size may be approximately summarized as the following:

1. *Gravel:* particles smaller than 3 in. and larger than $\frac{1}{4}$ in.
2. *Sand:* particles smaller than No. 4 sieve (approximately $\frac{1}{4}$ in.) and larger than No. 200 sieve. Particles smaller than No. 200 sieve are not visible to the naked eye.
3. *Silt:* particles smaller than 0.02 mm and larger than 0.002 mm.
4. *Clay:* particles smaller than 0.002 mm.

Many soils, however, consist of mixtures of particles of different size: silty sand, sandy clay, etc.

Sands have a specific gravity of closely 2.65. Clays are heavier, with specific gravities from about 2.70 to 2.9. The specific gravity is the weight of the material compared with the weight of the same volume of water.

## 4.10   *moisture content*

Much of the engineering behavior of soils is governed by moisture content. Clays shrink with loss of moisture and expand as they take up moisture. But unlike other types of soils, clays are relatively impermeable to moisture. Because of low permeability, clays may require months or years to attain new moisture equilibrium conditions. The erection of a building may alter moisture conditions in its subsoil. Slabs on grade, for example, interfere with the natural transfer of moisture into and from the ground, and may be subject to vertical movements from soil pressures. The clay at the edge of such a slab may become drier than the clay beneath the slab, and when this happens the soil may not support the slab at its periphery. Similarly, a plumbing leak beneath such a slab may produce heaving.

All soils are a skeleton of soil particles with intervening voids filled with water or air or both. A saturated soil is the special case where all the voids are filled with water.

The water in soils may be gravitational, capillary, or hygroscopic. The surface below which water is continuous is called the *water table*. The water below this level and water percolating down to it, is *gravitational water*. Wells must be drilled below the water table if water is to be pumped continuously. Above the water table capillary action draws water upward into the voids in the soil. Such water above the water table is *capillary water*. Finally, molecular attraction causes water molecules to be attached to the surface of the grains of the soil, and such water is called *hygroscopic water*. The amount of both hygroscopic and capillary water will increase with decreasing grain size, and therefore will be greatest in clays.

If a soil sample is air-dried, both the gravitational and capillary water will evaporate. Hygroscopic water, however, remains in an air-dried soil. Hygroscopic water is removed by heating the sample at 105°C, just over the boiling point of water, for 24 hr. The water content of a soil sample thus can be determined by weighing the sample before and after such heating. The water content is defined as the weight ratio of water to oven-dried soil:

$$\text{water content} = \frac{W_w}{W_s}$$

This is expressed as a percentage. The water content of a clay may exceed 100 per cent.

To convert from dry to wet densities, consider the following example.

A granular soil weighs 110 pounds per cubic foot (lb per ft³) when oven-dried. A test sample from the site shows a water content of 11 per cent. What is the wet density of this soil?

weight of water $= 0.11 \times 110 = 12.1$ lb

wet density $=$ soil plus water $= 110 + 12.1 = 122.1$ lb per ft³

The *void ratio* can also be determined if the specific gravity is known from a specific gravity test. The following is an example.

The same soil has a specific gravity (s.g.) of 2.65. What is the void ratio of this soil?

$$\frac{\text{absolute volume of soil}}{\text{solids per cubic foot}} = \frac{\text{wt solids}}{\text{wt water/ft}^3 \times \text{s.g.}} = \frac{110}{62.4 \times 2.65} = 0.665 \text{ ft}^3$$

volume of voids $= 1 - 0.665 = 0.335$ ft³

Therefore, the porosity of this soil is 0.335/1,000, or 0.335, as a ratio. The void ratio of voids to solid matter $= 0.335/0.665 = 0.504$.

## 4.11  plasticity

One of the most distinguishing characteristics of clays is their *plasticity*. The degree of plasticity depends on the water content of the clay. If a clay sample is mixed with considerable water, the mixture will act like a fluid and has no shear resistance. If the water content is steadily reduced, a water content will be reached at which the soil will exhibit some shear strength. The maximum water content at which shear strength first appears is called the *liquid limit*, $w_l$. The liquid limit is the water content separating the liquid and plastic conditions.

If the water content of the soil is further reduced below the liquid limit, the shear strength will increase and the plasticity will decrease. The soil volume will decrease also, of course. Finally, with a continued decrease in water content a point will be reached at which strain will produce rupture

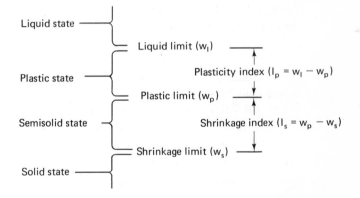

Fig. 4.11  Consistency limits (also called Atterberg limits).

instead of plasticity. This point is called the *plastic limit*, $w_p$, and may be considered to be the water content that separates the plastic and semisolid states. Reduced water contents below the plastic limit will eventually bring the soil to the least water content for complete saturation, the *shrinkage limit*, $w_s$. Below $w_s$ there is no more shrinkage. The shrinkage limit may be considered to be the water content that separates the semisolid and solid states. See Fig. 4.11 for a summary diagram of these limits, which are sometimes referred to as *consistency limits*.

The range of water contents across which the soil is plastic is termed the

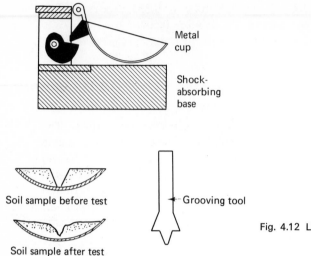

Metal cup

Shock-absorbing base

Soil sample before test

Grooving tool

Soil sample after test

Fig. 4.12 Liquid-limit apparatus.

*plasticity index, $I_p$.* $I_p$ is the difference between the liquid and plastic limits:

$$I_p = w_l - w_p$$

Similarly, the *shrinkage index $I_s$* is the difference between the plastic and shrinkage limits:

$$I_s = w_p - w_s$$

The plastic limit is determined by the smallest value of moisture content at which the soil can be rolled into a $\frac{1}{8}$-in.-diameter thread without crumbling, that is, the point at which brittle failure commences.

The liquid-limit apparatus is shown in Fig. 4.12. The purpose of the

Fig. 4.13 Determination of liquid limit from graphed data.

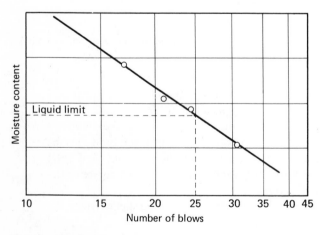

Moisture content

Liquid limit

10      15      20    25    30    35    40  45
Number of blows

liquid-limit test is to find the point at which the soil begins to exhibit shear strength. This is the moisture content at which a sample placed in the liquid-limit cup, leveled, and separated into two halves by a standard grooving tool will flow together for $\frac{1}{2}$ in. of length when jarred 25 times by the apparatus through a crank turned at 2 revolutions per second. The liquid limit is found from a series of these tests, plotting the number of jars or blows versus water content on semilogarithmic paper. A straight line results, as in Fig. 4.13. The 25-blow point on the line is taken as the liquid limit.

The following table summarizes plasticity in soils:

| Soil Type | Plasticity | Plasticity Index |
|---|---|---|
| silt | not plastic | 0 |
| clayey silt | slight | 1–5 |
| clay and silt | medium | 10–20 |
| silty clay | high | 20–40 |
| clay | very high | 40 or more |

Note that silt is a noncohesive soil. A soil with a high plasticity index and high liquid limit will shrink and swell excessively.

## 4.12 freezing in soils

In colder latitudes the depth of frost penetration into the ground tends to increase with the *freezing index*. This index is the number of degree-days of frost during the year. A degree-day of freezing results when the mean outside air temperature for 1 day is 1°F below 32°F. Hence if the average temperature for a certain day is −3°F, this day contributes 35 degree-days of frost.

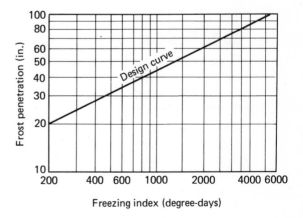

Fig. 4.14 Frost penetration as related to freezing index.

Figure 4.14 shows the relationship between the depth of frost penetration and the freezing index. Freezing index is the major influence on frost penetration, although other factors have their effects. Snow cover or straw cover act as ground insulation and will have a marked effect on frost penetration. On a certain freshwater lake at latitude 59°N (Lake Isabelle, Fort Churchill Army Base) it was found that the total of the thickness of the ice layer on the lake plus the snow cover always summed very closely to 5 ft.

Frost may produce remarkable frost heaving. When water freezes to ice there is a linear expansion of 9.4 per cent. Since frost heaves as great as a few feet have been known, volume expansion cannot account for such large movements.

As the temperature drops at the end of the year, the plane of freezing (32°) slowly penetrates the soil. At this plane, water turns to ice. Such freezing is actually a drying action. Water in the unfrozen soil below the freezing plane will move by capillary action toward the freezing plane, that is, from moist soil to dry soil, thus increasing the volume expansion by contributing to the size of the ice lens in the soil. If the soil is moist before the onset of winter, such an ice lens can be continuously fed with water and grow to enormous size. Such frost heaves will occur only in moist fine-grained soils such as clays, and especially in silts. They are prevented either by drainage or by replacing the fine-grained soil with a coarser granular material, usually gravel. They do not occur under heated buildings.

## 4.13  settlement of foundations

Strain accompanies stress. For many construction materials strain occurs simultaneously with stress. There are, however, many construction materials that have a considerable time lag between the application of the stress and the resulting strain. For such materials, the strain will continue to increase for days or weeks after application of the load. This is true for many plastics and rubbers (Fig. 1.6 shows such a time lag), for asphalts, and for fine-grained soils and sands. In the case of soils, this time lag for the appearance of the strain is called *consolidation*.

Only modest loads are applied to soils by foundations, approximately 1 ton/ft². For such low stresses, there can be little deformation of the solid material of the soil. The change in volume that gives the apparent strain must be due to a decrease in the volume of voids in the soil. In clays of low permeability and high moisture content, water must flow out of the stressed soil if the soil is to consolidate, and the low permeability will mean that consolidation may carry on for months. In extreme cases, settlement of historic heavy stone structures in Europe has continued for centuries.

Figure 4.15 shows two graphs, the first of which is a typical compression–time curve for a backfilled sand without clay. Theoretically at any rate,

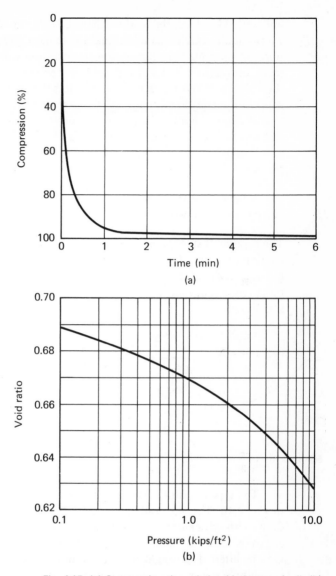

Fig. 4.15 (a) Compression-time relationship for a sand soil; (b) void ratio of sand as a function of soil pressure.

this sand will consolidate infinitely at an extremely slow rate. The second graph does not apply to the same sand as the first graph, but is a typical graph of void ratio against pressure.

Consider a saturated fine-grained clay under a pressure $P_1$ due to the weight of soil above it, with a void ratio $e$. Now impose the additional weight of a structure on this soil so that the pressure increases suddenly to $P_2$. The void ratio is still $e$, since time is required to alter this in a relatively impermeable soil. Therefore, the load on the soil is still $P_1$. The additional load $P_2 - P_1$ is initially carried as additional hydrostatic pressure by the water in the pores of the soil. This increased water pressure will cause water to permeate from the loaded soil to soil areas of lower pressure. As water permeates away, the volume of voids is reduced and the solid material of the soil begins to support the additional load. As the increased hydrostatic pressure falls with loss of water, the rate of loss of water also falls, and the volume change in the soil will follow the trend of Fig. 4.15(a).

It may be seen that the problem of settlement of a foundation is a double problem:

1. What will be the amount of the total settlement?
2. At what rate will this total settlement be achieved?

The making of such predictions is the job of the soils engineer. The allowable bearing capacities for various soil and rock classifications are given in many building codes. Nevertheless, soil conditions are so variable and uncertain that often a soils specialist must analyze soil samples from the site for foundation design.

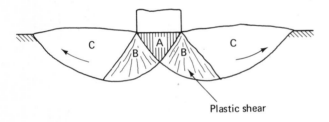

Fig. 4.16 Mechanism of footing settlement.

The type of failure exhibited by a plastic soil is a shear failure, as shown in Fig. 4.16. The wedge-shaped area $A$ beneath the footing moves vertically down with settlement. This movement gives a lateral thrust to areas $B$, which are called *zones of radial shear*. The large strains in zones $B$ cause the zones $C$ to be lifted. The weight of material in zones $C$ resists the lifting forces and is the force that reacts against footing settlement. Bearing capacity therefore depends on the weight of soil thus resisting settlement and the friction characteristics of the soil. These soil characteristics can be determined by laboratory investigations.

Settlement of a foundation may have any of the following causes:

1. Static load of the structure.
2. Moving loads, usually causing failure of road or airport pavements.
3. Changes in the moisture content of the soil.
4. Undermining by adjacent excavation.

In a *granular soil* such as sand, the load capacity of a pile is given by skin friction of the soil against the pile. Skin friction increases with depth of pile penetration. The pile-driving operation vibrates the granular soil and compacts it, and as a result there may be settlement of the ground surface.

In a *cohesive soil* such as clay, the action of a pile is entirely different. The pile is driven in a relatively short time interval within which there is insufficient time for soil water to permeate from the region of the pile. As explained above, the soil stress caused by the pile must be carried by hydrostatic pressure of the water in the voids. In time, pore water will dissipate, the soil around the pile will gain in shear strength, and the pile will gain in load-carrying capacity. Since a cohesive soil cannot be consolidated in the same way as a granular soil, the pile-driving operation will cause the ground surface to rise. It may even cause an adjacent pile to rise.

## PERTINENT ASTM SPECIFICATIONS

C99-52 (Part 12) Modulus of Rupture of Natural Building Stone
C170-50 (Part 12) Compressive Strength of Natural Building Stone
C119-50 (Part 12) Terms Relating to Natural Building Stone

The following soils specifications are all in Part 2:

D653-67 Terms and Definitions Relating to Soil and Rock Mechanics
D2487-66T Classification of Soils for Engineering Purposes
D2488-66T Description of Soils (Visual-Manual Procedure)
D2113-62T Diamond Core Drilling for Site Investigation
D1586-67 Penetration Test and Split-Barrel Sampling of Soils
D1587-67 Thin-Walled Tube Sampling of Soils
D422-63 Grain Size Analysis of Soils
D1140-54 Amount of Material in Soils Finer Than the No. 200 Sieve
D1452-65 Soil Investigation and Sampling by Auger Borings
D854-58 Specific Gravity of Soils
D2216-66 Laboratory Determination of Moisture Content of Soil
D698-66T Moisture-Density Relations of Soils Using 5.5-lb. Rammer and 12-in. Drop
D1557-66T Moisture-Density Relations of Soils Using 10-lb. Rammer and 18-in. Drop
D423-66 Liquid Limit of Soils
D424-59 Plastic Limit and Plasticity Index of Soils
D427-61 Shrinkage Factors in Soils
D2166-66 Compressive Strength, Unconfined, of Cohesive Soils
D2664-67 Triaxial Compressive Strength of Undrained Rock Core Specimens
D1194-57 Bearing Capacity of Soil for Static Load on Spread Footings

## QUESTIONS

1  Explain the formation of igneous, sedimentary, and metamorphic rock deposits.

2  Why is the cost of stone proportional to hardness?

3  What are some of the reasons why solid stone walls have fallen into disfavor?

4  What is the meaning of (a) porosity; (b) permeability?

5  What is dimension stone?

6  Explain the terms cornice, quoin, and belt course.

7  Differentiate between artificial stone and synthetic stone.

8  What is the difference between hot lime and slaked lime?

9  Why is iron oxide objectionable in window glass?

10  What is meant by a rubble wall, a random ashlar wall, and a coursed ashlar wall?

11  A pile is partially driven into a cohesive soil, and then the driving of the pile is interrupted for a month by a labor strike. The driving of the pile is then completed. What effect will the delay have on the pile-driving operation?

12  Differentiate between residual and transported soils.

13  What characteristic differentiates a silt from a clay?

14  Differentiate among gravitational, capillary, and hygroscopic water.

15  If a granular soil with 13 per cent water content weighs 118 lb per ft$^3$ when oven-dried, what is its wet density?

16  Explain plastic limit, shrinkage limit, and liquid limit of soils.

17  Define plasticity index and shrinkage index.

18  Briefly summarize the method of the liquid limit test for soils.

19  What is a freezing index?

20  Explain why an impermeable clay is slow to consolidate.

21  State the four possible causes of foundation settlement.

22  In plotting a grain-size distribution curve, why is a logarithmic scale used instead of a linear scale?

23  Find the liquid limit for a soil from the following data:

| Water Content (%) | Number of Blows |
|---|---|
| $38\frac{1}{2}$ | 3 |
| $37\frac{1}{2}$ | 8 |
| 36 | 29 |
| 35 | 40 |

The words "cement" and "concrete" are often confused in conversation, as in referring to a concrete sidewalk as a "cement sidewalk." Cements, whether *organic*, such as rubber cement, or *inorganic*, such as the portland cement used in concrete, are adhesive materials. In a concrete, the cement must coat the surface of all the particles of the aggregate in order to bind the whole into a monolithic mass. The bulk material or filler material, usually crushed stone, gravel, or sand, is called *aggregate*.

## 5.1 manufacture of portland cement

The most important of the inorganic cements used in construction is *portland cement*. This is a silicate cement, and is produced in several types.

Almost all inorganic cements and mortars harden by taking up water or carbon dioxide in chemical reactions. The ancient Romans used a brick

**CEMENTS AND CONCRETE**

**5**

mortar resembling our own sand-lime mortar. It was a mixture of quicklime and burned clay in the form of crushed brick. Mortars, ancient or modern, harden very slowly as the lime combines with carbon dioxide in the air to form a rock-hard calcium carbonate:

$$CaO + CO_2 \longrightarrow CaCO_3$$

Portland cement is a more complex mixture of chemicals and has a more complex chemistry of hardening. Compared to lime, portland cement hardens quickly and attains much higher strengths after hardening. These effects are chiefly due to the presence of tricalcium silicate in the cement, a compound not usually found in lime mixtures.

The raw materials for portland cement are clay or shale, and limestone, $CaCO_3$. Occasionally chalk, oyster shells, or other calcium carbonates may be substituted for limestone. Clays are formed by the weathering of granites. Chemically they are complex aluminum silicates. Approximately 4 parts of limestone to 1 part of clay is used for the raw feed to be processed into portland cement. A small amount of iron oxide must be added as a flux to reduce the temperature at which the raw materials combine.

Pure limestone without any dolomite ($MgCO_3$) is rarely found. Dolomitic limestones, which are about half limestone and half dolomite, are unsuited to the manufacture of cement. Magnesia does not combine with the acid silica and remains as free magnesia in the finished portland cement. If free magnesia, or even free lime, were present in the cement, it would "slake" or combine with water to produce $Mg(OH)_2$. This reaction results in an increase in volume, and the effect would be to crack and disintegrate the concrete in which the cement was used. Expansion due to magnesia is more dangerous than expansion due to lime because its development is quite slow and its first effects would not appear until the passage of a few years. For this reason specifications for portland cement limit the magnesia content to a maximum of 5 per cent. There are always limited amounts of magnesia in any portland cement, and it is this material that gives cement its gray-green color.

A typical raw feed by weight for a portland cement might be (in per cent):

| | |
|---|---|
| $SiO_2$ | 15.5 |
| $Al_2O_3$ | 2.5 |
| $Fe_2O_3$ | 2.0 |
| $CaO$ | 42.0 |
| $MgO$ | 2.5 |
| $CO_2$ | 35.5 |
| | 100.0 |

Here the $CO_2$ is combined as $MgCO_3$ and $CaCO_3$. The raw limestone, clay, and iron oxide must be ground to pass a 200-mesh sieve in a ball mill.

The raw materials are fed into the cement kiln for burning, either dry or, more commonly, as a slurry with water.

The *cement kiln* is an impressive piece of equipment. It is a long cylinder, not less than 11 ft in diameter and not less than 400 ft long, with a slope of about $\frac{1}{2}$ in./ft toward the discharge end. It is lined with firebrick 9 in. thick. The kiln is supported on heavy rollers and rotated at about 0.6 rpm by a large electric motor of 125 hp or more. A standby diesel engine is necessary because if the kiln stops rotating, its great weight and length will cause it to warp and sag. The kiln is fired by a large burner at the discharge end. An interior view of an 11-ft-diameter cement kiln is shown in Fig. 5.1.

Fig. 5.1 The interior of a cement kiln under repair. The globules of clinker attached to the interior of the kiln are characteristic.

The raw kiln feed is metered into the kiln at the high end and is tumbled down to the discharge end in a time of a few hours. The tumbling action produces a product in pellet form. The raw feed is progressively heated as it moves down the length of the kiln, and the following sequence of processes occurs:

1. Water is driven off.
2. At 630°F, $MgCO_3$ decomposes to $MgO$ and carbon dioxide.
3. At 1630°F, $CaCO_3$ decomposes to $CaO$ and carbon dioxide.

4. Finally, in the "burning zone" of the kiln, which is about the last 70 ft, the components undergo a solid-state chemical reaction at a temperature of about 2700°F, which produces calcium silicates. The reaction that is the most difficult to complete is the combination of the last trace of lime with previously formed dicalcium silicate to form tricalcium silicate. The amount of lime remaining uncombined is usually less than 1 per cent.

The product of the kiln is a pelletized black-green slag called *clinker*. This is dropped through the floor of the firing hood that encloses the end of the kiln and is conveyed through a cooler.

The final operation at the cement mill is to convert the clinker into cement. The clinker is crushed, about 3 per cent gypsum is added, and the product is reduced by grinding to 325 mesh.

## 5.2 types of portland cement

Portland cements are composed of four principle chemical compounds plus the added gypsum:

1. Tricalcium silicate, $3CaO \cdot SiO_2$, abbreviated $C_3S$.
2. Dicalcium silicate, $2CaO \cdot SiO_2$, abbreviated $C_2S$.
3. Tricalcium aluminate, $3CaO \cdot Al_2O_3$, abbreviated $C_3A$.
4. Tetracalcium aluminoferrite, $4CaO \cdot Al_2O_3 \cdot Fe_2O_3$, abbreviated $C_4AF$.
5. Gypsum, which is calcium sulfate, $CaSO_4$.

The percentages of the four major constituents can be varied to provide cements with a range of characteristics. The strength of portland cement is controlled by the amount of the two calcium silicates, which together constitute about 70 per cent of the cement.

Three types of portland cement are commonly used: normal portland cement (type I), high-early-strength cement (type III), and sulfate-resistant cement (type V). Other types include low-heat cement (type IV), modified portland cement (type II), and oil-well cement. These cements have the typical compositions in per cent indicated in Fig. 5.2.

All four chemical constituents in portland cement take up water in different amounts and at different rates during the setting of the cement, and by adjusting the relative amounts of these constituents the properties of the cement can be adjusted. When cements harden or cure, whether they are inorganic or organic cements, heat is produced. In the case of organic adhesives, this heat is called *exotherm*. For hydraulic cements such as portland cement, this heat is called the *heat of hydration*. The faster the cement cures, the higher will be the exothermic temperature. Rapid-setting cements, such as high early strength, develop high temperatures while setting; slower-setting types such as type II produce lower temperatures.

| Compound | Type I | Type II | Type III | Type IV | Type V |
|----------|--------|---------|----------|---------|--------|
| $C_3S$ | 45 | 44 | 53 | 28 | 38 |
| $C_2S$ | 27 | 31 | 19 | 49 | 43 |
| $C_3A$ | 11 | 5 | 10 | 4 | 4 |
| $C_4AF$ | 8 | 13 | 10 | 12 | 8 |
| | Percent passing 325 mesh sieve | | | | |
| | 90.7 | 94.7 | 99.5 | 93.1 | 93.2 |

Fig. 5.2 Analysis and sieve fineness for five types of Portland cement.

The setting rate of both tricalcium and dicalcium silicate is low, and if these two were the only constituents, the cement would remain plastic for several hours after mixing with water. The opposite occurs with tricalcium aluminate. This constituent develops a "flash set" when mixed with water, with the release of considerable heat. Although the proportion of $C_3A$ in any type of cement is not high, as shown in Fig. 5.2, nevertheless its presence would lead to flash setting of the whole concrete mass. This explains the necessity for adding a small amount of gypsum to cements. Gypsum prevents such flash setting and regulates the setting time. It does this by quickly combining with $C_3A$ to form a needle-like constituent, calcium sulfoaluminate, which contains a considerable amount of chemically bound water or "water of crystallization."

Tricalcium silicate releases twice as much heat of hydration per unit weight as dicalcium silicate. Tricalcium aluminate releases three times as much. Type IV cement, being low in $C_3S$ and $C_3A$, will generate less heat of hydration than the other types of portland cement.

**Type I: Normal portland cement.** Type I cement is used for general construction work when no special properties are required of the cement.

**Type II: Modified portland cement.** Type II cement generates heat at a less rapid rate than normal portland cement. Hence it is better suited to more massive concrete pours, where the large volume of the pour and relatively lesser surface area reduce the cooling capacity of the pour. This cement also has a better resistance to the attack of sulfates in soils.

**Type III: High-early-strength cement.** Type III cement is used when high strength is needed very quickly in the cement. A high early concrete strength would be required if forms must be removed as soon as possible, or in cold-weather operations, when it is desirable to reduce the time needed to protect the concrete from freezing.

**Type IV: Low-heat cement.** Low-heat cement is used in very massive concrete pours such as dams. This type is formulated for minimum heat and slow heat generation and is used when there is danger of cracking of the concrete due to expansion and subsequent contraction from temperature changes during setting of the concrete.

**Type V: Sulfate-resistant cement.** Sulfates are often found in soils and water. Such sulfates attack cement by reacting with calcium hydroxide in the cement to produce first gypsum, $CaSO_4$, and then a water-rich calcium sulfoaluminate. This chemical action is harmful because it produces a volume expansion. To prevent such action, sulfate-resistant cement is low in $C_3A$.

A comparison of the properties of the several types of cement is given in the following table. The resistance to chemical attack may be important for concrete floors in industrial buildings. Such products as milk (lactic acid), vinegar (acetic acid), and blood (a saline solution) may be damaging to ordinary concrete.

### CHARACTERISTICS OF HYDRAULIC CEMENTS

| | Rate of Strength Development | Heat Evolution | Drying Shrinkage | Resistance to Cracking | Resistance to Chemical Attack |
|---|---|---|---|---|---|
| normal portland | medium | medium | medium | medium | limited |
| low heat | low | low | greater | high | fair |
| rapid hardening | high | high | medium | low | low |
| sulfate resisting | medium | medium | medium | medium | high |
| high alumina | very high | very high | medium | low | very high |
| pozzolan cements | low | low | medium | high | high |

## 5.3 hydration of portland

## cement

Portland cement, like any hydraulic cement, sets and hardens by taking up water in complex chemical reactions. This chemical hardening process is called *hydration*. Provided that a source of water is present at all times— water vapor in the air is a suitable source of water—portland cement will

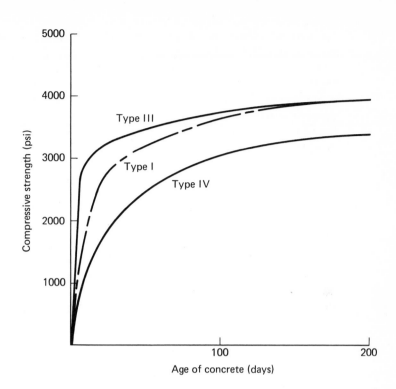

Fig. 5.3 Representative comparison of the development of strength in three types of concrete.

continue to harden over a period of months, as may be seen from an examination of Fig. 5.3. The first reaction to be completed is the combination of gypsum with $C_3A$ and water to form calcium sulfoaluminate; this chemical action is completed within 24 hr. The other compounds in cement will hydrate by combining with about 25 per cent of their weight of water. The amount of water added to cement exceeds this quantity, however.

$C_3S$ hydrates in the following reaction:

$$2C_3S + 6H_2O \longrightarrow 3CaO \cdot 2SiO_2 \cdot 3H_2O + 3Ca(OH)_2$$

while $C_2S$ hydrates similarly:

$$2C_2S + 4H_2O \longrightarrow 3CaO \cdot 2SiO_2 \cdot 3H_2O + Ca(OH)_2$$

Thus the final products of hydration of both calcium silicates are $C_3S_2H_3$, which is a compound called tobermorite, and lime or calcium hydroxide. Note that $C_3S$ releases in its hydration three times as much calcium hydroxide. $C_3A$ hydrates without taking up or releasing lime. Some of the lime released by the hydration of calcium silicates is taken up in the hydration of $C_4AF$.

The workability of cement paste does not change for some time after water is added. In this apparently dormant period, there are rapid and complex chemical reactions, including the gypsum and $C_3S$ reactions already mentioned. Setting of the cement occurs within a few hours, but hardening is not completed for months, although there is no fundamental difference between setting and hardening. *Setting* is the disappearance of plasticity in the cement paste which begins when water is added; *hardening* is the development of strength over an extended period of time as cement and water combine. Rapid setting is therefore not the same as rapid hardening. Gypsum retards setting, and the use of hot water accelerates setting. The finer the cement powder is ground, the greater the amount of cement surface exposed to water and the faster the rate of setting.

Rapid setting and high early strength are provided by an increase in $C_3S$, as shown in the analysis of high-early-strength cement. Fast-setting cements, however, release their heat of hydration rapidly and thus produce elevated temperatures in the concrete. Hence low-heat, type IV cement is low in $C_3S$.

The ultimate compressive strength developed in a portland cement results from the hydration of the two calcium silicates. $C_3A$ and $C_4AF$ contribute little to the strength of concrete, although they reach their maximum strength levels in a short period of time.

The heats of hydration of the four cement compounds are these (in calories per gram):

| | |
|------|-----|
| $C_3S$ | 120 |
| $C_2S$ | 62 |
| $C_3A$ | 207 |
| $C_4AF$ | 100 |

Low-heat, type IV cement is low in $C_3S$ and $C_3A$, the two compounds that generate the most heat.

## 5.4  high-pressure steam curing

The curing of concrete by high-pressure steam is also referred to as *autoclaving*, since the process must be carried out in a pressure vessel called an autoclave. Autoclaving was first employed for curing sand-lime brick, and is still used for this purpose, producing a reaction between the silica sand and the lime to form calcium silicates of high strength. Both standard and lightweight concrete products are stream-cured. The method has the following advantages for concrete products:

1. High early strength is developed. The normal 28-day strength by atmo-

spheric curing may be obtained in 24 hr with high-pressure steam.

2. Improved resistance to sulfates and some other types of chemical attack.
3. Improved resistance to freezing and thawing.
4. Reduced curing shrinkage. This is especially important in the case of lightweight concrete products with their higher curing shrinkage.

High-pressure steam curing is most effective when silica ground to the fineness of the cement is added. The amount of silica flour should be between 40 and 70 per cent of the weight of the cement. The silica reacts with the lime released by hydration of $C_3S$. Cements rich in $C_3S$ develop higher strength than those with high $C_2S$ in steam curing. Steam curing, however, can rarely be used for reinforced concrete products, since the bond strength to the reinforcement is reduced.

## 5.5 variations of portland cement

*Portland blast furnace cement* is made by grinding together portland cement clinker and granulated blast furnace slag. This slag is a mixture of lime, silica, and alumina, and thus has the same oxides as portland cement, although not in the same proportions.

*White portland cement* is a normal portland cement made from materials selected for a pure white color. Such a cement may be made from chalk or kaolin (china clay) with very little iron oxide or manganese oxide. This cement usually has a lower strength than other portland cements, but since white cement is largely used as a facing against standard concretes, strength is less critical.

*Masonry cement* provides a mortar superior to that made with standard portland cements. This cement is made from a mixture of normal portland cement clinker, high-calcium limestone, and an air-entraining agent. Masonry cements give a more plastic mortar with less shrinkage. The strength of the mortar will be lower, but this is generally an advantage.

*Aluminous cements*, also called *lumnite cement* and *Ciment Fondu*, are high-alumina cements used in the mixing of refractory concretes for lining furnaces, chimneys, and boilers. Since the alumina content is high, the silica content is low, ranging from 3 to 11 per cent.

Lumnite cement is more expensive than portland cement, but has uses in construction. In 24 hr it develops the same strength that portland cement develops in 28 days. A lumnite concrete can therefore be put into service in a very short time. The rapid hardening rate develops elevated temperatures which are useful in cold-weather pouring of concrete, although protection against freezing is necessary until temperature elevation begins. If the temperature of the concrete becomes too high, there is some loss of strength in the concrete. Because of this rapid hardening rate, lumnite cement must not be poured in heavy sections.

Hydration of lumnite cement releases free alumina instead of calcium hydroxide. This explains the corrosion resistance of lumnite cement, which is not attacked by sulfates or seawater.

*Portland–pozzolan cements* are blended mixtures of portland cement and pozzolans. Pozzolans are natural or artificial materials containing reactive silica. These siliceous materials combine in the presence of water with the lime released by the hydration of cement to form cementing calcium silicates. A wide range of materials comprise possible pozzolans: calcined clays and shales, blast furnace slags, crushed brick, pumicite, and flyash from large steam-generating plants that are coal-fired.

Pozzolan cements gain strength very slowly and require a long curing period. Hence they are low-heat cements of importance for massive concrete pours. Their favorable effects upon concrete include the following:

1. Improved workability.
2. Low heat of hydration.
3. Improved watertightness.
4. Resistance to sulfate attack.
5. Reduced cost.
6. Reduced alkali-aggregate reaction (see Section 5.7).

## 5.6 concrete

The cement is the binder for the aggregate in concrete. Among the materials used for aggregate are gravel, sand, crushed stone, crushed slag, and expanded minerals such as perlite,. vermiculite, and clays. While the strength of a concrete is chiefly determined by the strength of the cement, a weak aggregate such as vermiculite cannot produce a high-strength concrete. Clearly no concrete can have a compressive strength exceeding that of its aggregate; however, the compressive strength of most aggregates greatly exceeds that of most concrete.

In a standard concrete about 75 per cent of the volume of the mix is occupied by aggregate materials. Since the cement is the more expensive component of the concrete, the amount of cement must be minimized consistent with the required strength and quality of the concrete. The minimum cement requirement is realized by suitably grading the aggregate so that small particles can fill the voids between the larger pieces of aggregate. Each aggregate particle should be completely embedded in cement paste, except for voids and entrained air in the concrete. The hardened concrete will weight about 150 lb per ft$^3$, except for lightweight concretes and special heavy concretes. Most concretes are batched to provide a specified minimum compressive strength after 28 days. This will usually lie within the range 2000–6000 psi.

The tensile strength of concrete is low but is proportional to the compressive strength. The modulus of elasticity of concrete is about 1000 times the ultimate compressive strength, although strain is not proportional to stress in concretes.

## 5.7 stone aggregate

The particles used for aggregate in standard concretes must be hard and strong and of a suitable shape. Angular and sharp pieces require more cement and fine material to make a good concrete. Flat or flaked pieces can, in addition, collect water or voids under their flat surfaces.

Harmful materials in aggregate may be of three possible types:

1. Weak or unsound particles, such as some shales or iron sulfides.
2. Particles with coatings that prevent the development of a good bond with the cement paste.
3. Impurities that interfere with the hydration of cement.

A fourth type of harmful aggregate material is occasionally found in certain areas, causing an alkali-aggregate reaction. In this reaction a cement with an unusually high content of sodium and potassium oxides attacks susceptible aggregates containing silica, resulting in disruptive pressures in the concrete after a period of time.

Surface coatings that can interfere with the bond between cement and aggregate include clay, silt, crusher dust, and organic impurities that can affect the setting of the cement paste. The silt test determines the amount of clay and silt present in sand (Fig. 5.4). In this test, sand to a depth of 2 in. is placed in a quart jar and clean water added to make the jar three-fourths full. The top is screwed on the jar and the contents are well shaken. The jar then sits for several hours. As the materials settle, the heavier sand deposits first and the fine material is deposited last on top of the sand. When the water finally becomes clear, the depth of the silt deposit is measured. If it exceeds $\frac{1}{8}$ in., the aggregate contains an excessive amount of fine material and is not suitable for concrete unless washed.

A colorimeter test for organic material in sand (Fig. 5.5) uses a 12-oz prescription bottle filled to the $4\frac{1}{2}$-oz mark with a sample of the sand to be tested. A 3 per cent solution of sodium hydroxide in water is added to fill the bottle to the 7-oz mark. The contents are well shaken, then left to stand for 24 hours. If the liquid, originally clear, becomes colored, this coloration indicates the presence of organic matter. A color range from

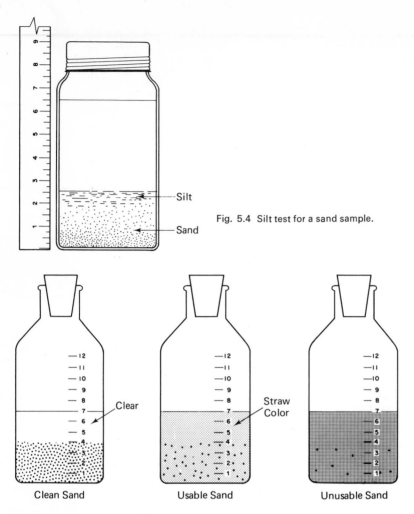

Fig. 5.4 Silt test for a sand sample.

Fig. 5.5 Colorimeter test for a sand sample.

light to dark straw indicates that the organic content is not seriously high. Darker colors normally indicate the need for washing the sand, although dark colors can also be caused by harmless iron compounds.

Sieve analyses test the gradation of fine and coarse aggregates. A sieve series of 6, 3, 1½, and ⅜ in. plus No. 4 is used for coarse aggregate, and Nos. 4, 8, 16, 30, 50, and 100 are used for fine aggregate. A No. 50 sieve has 50 openings per lineal inch. The series of sieves divide a sample of the aggregate into fractions each consisting of particles of approximately the same size. All sieves are nested in frames (Fig. 5.6) with the coarsest sieve on top and the finest at the bottom.

A 1000-g sample of fine aggregate is passed through the sieves. The amount retained on each sieve is weighed and calculated as a percentage.

Fig. 5.6 Nest of sieves for concrete aggregate.

The results should fall within the following ranges:

| Sieve Size | Per cent Retained (Cumulative) |
|---|---|
| 4 | 0– 5 |
| 8 | 10–20 |
| 16 | 20–40 |
| 30 | 40–70 |
| 50 | 70–88 |
| 100 | 92–98 |

As an example, suppose that a sample gives the following quantities on each sieve:

| Sieve | Grams | Per Cent Each Sieve | Per Cent Cumulative |
|---|---|---|---|
| 4 | 31 | 3.1 | 3.1 |
| 8 | 125 | 12.5 | 15.6 |
| 16 | 230 | 23.0 | 38.6 |
| 30 | 262 | 26.2 | 64.8 |
| 50 | 210 | 21.0 | 85.8 |
| 100 | 139 | 13.9 | 99.7 |
| | 997 | | 307.6 |

The *fineness modulus* is a single factor calculated from the sieve analysis, actually the weighted average sieve size. The fineness modulus is found by adding the cumulative percentages, which in this case total 307.6, and dividing by 100. Therefore, the fineness modulus of the sample is 3.08. The allowable fineness modulus should lie between 2.20 and 3.20, a lower modulus indicating fine sand, and a higher modulus a coarse sand.

## 5.8 lightweight aggregate

*Expanded shale or clay*, also called *haydite* or *herculite* (see Fig. 5.7), is produced by heating suitable clays. Gases within the material cause bloating. The particles pelletize in a range of sizes and are screened into fine, medium, and coarse material. The coarse size ranges from $\frac{3}{8}$ to $\frac{3}{4}$ in.

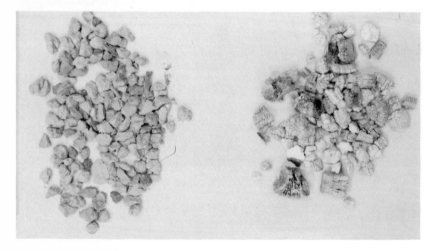

Fig. 5.7 Haydite and vermiculite.

Haydite is employed in concrete block, roof planks, and other precast shapes. It has the advantage of a weight reduction of about one-third, and being cellular offers some thermal insulation. Acoustic absorption is about double that for heavy concrete, although no type of concrete is actually an "acoustic" material. As with most lightweight concretes, the curing shrinkage is greater and more water is required than for standard aggregates. Since haydite absorbs water, it is prewetted with about half the required water before the cement and remaining water are added. This procedure prevents the waste of cement in the voids of these rough-surfaced aggregates, where it is assumed to be ineffective. Haydite aggregate is lightweight and therefore has a strong tendency to segregate, that is, to separate into coarse and fine fractions during handling.

*Vermiculite* (zonolite) is a lightweight material used either as a loose-fill thermal insulation or as a very lightweight aggregate for concrete products and plaster (Fig. 5.7). It is too soft and weak a material to be used in concretes that require strength. Vermiculite is a foliated material that expands under heat in an accordion-like fashion to give large particles of very low weight. It is used as an aggregate with or without sand. The $K$ factor for vermiculite concrete is about 1.0 Btuh/°F-ft$^2$-in.

*Perlite* is a volcanic material which when heated expands by the evolution of steam to form a light cellular material. When used as an aggregate for concrete it offers only limited strength with high shrinkage. Perlite is also an important ingredient in plaster mixes.

Fig. 5.8 An all-fuel prefabricated chimney designed by the author, insulated with perlite lightweight concrete.

*Foamed blast furnance slag* is expanded by exposing it to a water spray. That material expands by generation of steam.

Other lightweight concretes have been made with many types of materials, including diatomite, sintered flyash, sawdust, and wood particles. A cork aggregate produces a concrete that will float.

Figure 5.9 summarizes information on lightweight concretes. The modulus of elasticity of a lightweight concrete increases with its unit weight and compressive strength, and approximates 500 times the compressive strength of the concrete. The modulus of elasticity of a standard concrete is about 1000 times the compressive strength. Under similar conditions of loading a lightweight concrete member will deflect more than a standard concrete member.

Figure 5.9 shows that the curing shrinkage of lightweight concretes may be considerably larger than that of ordinary concrete. This shrinkage is

| Type of concrete | Compressive strength psi at 28 days | Weight (pcf) | | Bags cement per yd concrete |
| | | Loose Aggregate | Concrete | |
|---|---|---|---|---|
| Pumice | 300-2000 | 30-55 | 40-90 | 3-9 |
| Perlite | 100-800 | 5-15 | 25-60 | 3-8 |
| Vermiculite | 100-600 | 4-12 | 20-50 | 3-8 |
| Haydite | 600-6000 | 35-65 | 60-120 | 3-9 |
| Foamed slag | 400-5000 | 20-60 | 60-120 | 3-9 |
| Flyash | 600-6000 | 40-60 | 60-120 | 3-9 |

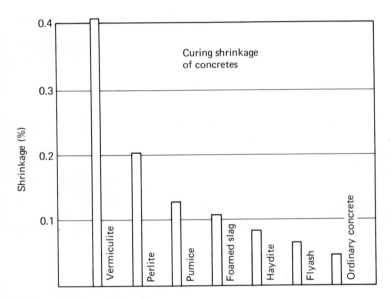

Fig. 5.9 Characteristics of lightweight concrete.

roughly inversely proportional to the unit weight of the aggregate. Even for ordinary concrete, a shrinkage of 0.5 per cent is a large amount, 1 in. in 16 ft. Fortunately, such shrinkage is partly compensated by the capacity of concrete, especially lightweight concretes, for plastic creep, which tends to heal any cracks that develop from such shrinkage. In an ordinary concrete the hard aggregate stiffens the mix against shrinkage. The principal factors that influence the relative curing shrinkage of concretes are these:

1. *Water–cement ratio:* one of the chief factors that influence shrinkage. If all other factors are the same, the shrinkage of a concrete will be roughly proportional to the water–cement ratio.
2. *Type of aggregate:* soft aggregate such as vermiculite will contribute

to shrinkage, whereas a crushed limestone may be considered as virtually dimensionally stable in the presence of water and other conditions.

3. *Proportion of fine aggregate:* the coarser the aggregate, the less the shrinkage.

4. *Setting accelerators:* setting accelerators such as calcium chloride and triethanolamine will increase curing shrinkage.

5. *Vibration:* vibrating the concrete appears to reduce shrinkage by about 20 per cent.

Regardless of aggregate size, an increase in the slump of concrete from 2 to 6 in. appears to increase the shrinkage by 25 per cent.

## 5.9   design of concrete mixes

The production of a serviceable and durable concrete at an economical cost requires careful control of the design and handling of the mix.

Fig. 5.10 Effect of water-cement ratio on compressive strength of concrete. Note that air entrainment reduces strength by 20 per cent.

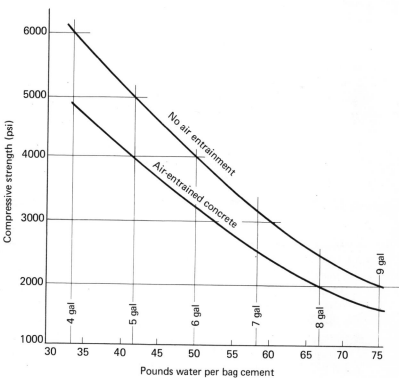

Virtually all concretes must meet a specification for compressive strength after 28 days. The major influence on the strength of a concrete is the water–cement ratio. The following other factors are usually of less importance in their influence on strength: strength of aggregate, cement-to-aggregate ratio, type of cement, temperature of concrete curing, and degree of compaction, that is, absence of voids. In addition to compressive strength, the maximum size of coarse aggregate must be decided, since this determines the proportion and grading of fine and coarse aggregate. This maximum size is governed by the thickness of the pour and the spacing of reinforcing bars.

The compressive strength of concrete is inversely proportional to the weight of water used per bag of cement. The objective in proportioning the water is the lowest water content that will give the required strength and at the same time produce a concrete sufficiently plastic to pour into the forms and to work successfully.

In proportioning water, allowance must be made for the fact that sand frequently contains a considerable amount of water. Deductions must be made for this circumstance. The water content of a sand sample is determined by weighing the sand before and after oven-drying. Figure 5.10 shows the probable minimum compressive strength of concretes after 28 days for various weights of water per bag of cement. One bag of cement is 94 lb. For concretes in thin or moderate sections and exposed to freezing air temperatures, the water–cement ratio should provide a strength not less than 3000 psi.

For small jobs, where the cost of the concrete and its performance are not critical, the following proportions may be used for the mix. The figures do not apply to lightweight aggregates or to air-entrained concrete, and it is assumed that the amount of water is controlled.

| Largest Aggregate | Sacks Cement per Yard of Concrete | Pounds Dry Sand per Sack of Cement | Pounds Gravel or Stone per Sack |
|---|---|---|---|
| 1 in. | 6.2 | 225 | 275 |
| 2 in. | 5.6 | 225 | 360 |

In order to calculate concrete batch quantities and water–cement ratios for larger jobs, where cost control and quality control become more significant, the following items of information about the aggregates are needed:

1. Specific gravity of the aggregates.
2. Unit weight of the aggregate in bulk form.

3. Gradation of aggregates.
4. Free moisture in the aggregate.

Figure 5.11 represents a concrete. The space or volume between the particles of aggregate is filled with cement paste. The total volume of concrete equals the volume of the cement paste plus the volume of the aggregate particles. The space occupied by the particles is their solid volume, not the loose volume of the particles as found in a pile of aggregate. In a loose pile there is considerable void space between particles and, in addition, surface moisture separates small aggregate particles from one another. This separation of small particles by water films is called *bulking*. The actual solid volume of aggregate, as explained below, is found from the weight and specific gravity. The volume of the cement paste is the sum of the volume of water plus cement plus trapped air. Air voids are frequently introduced into concrete by means of an air-entraining agent, but even without such an agent a cement paste will contain a small percentage of air voids.

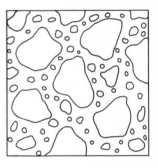

Fig. 5.11 Aggregate particles and cement paste.

There are many methods in use for the design of concrete mixes. Most of these methods use, or are modified from, the *absolute volume method*. This method is explained in publications of the Portland Cement Association and the American Concrete Institute. The method proportions the fine aggregate, coarse aggregate, cement, and water to provide the desired strength and other characteristics of the concrete and determines the volume of the fully compacted concrete produced by these materials.

The absolute volume method is based on the use of an absolute volume formula for the several components of the mix:

$$\text{absolute volume of a material} = \frac{\text{wt material}}{\text{specific gravity} \times \text{wt water/ft}^3}$$

The specific gravity of a material is the ratio of the unit weight of the material to the unit weight of water. The weight of 1 ft³ of water is 62.4 lb. Therefore, a material with a specific gravity of 2.0 would wight 124.8 lb per ft³.

## TABLE OF SPECIFIC GRAVITIES

| | Average Specific Gravity | Range of Specific Gravity |
|---|---|---|
| water | 1.0 | |
| cement | 3.15 | |
| sand or gravel | 2.65 | 2.5–2.8 |
| limestone | 2.65 | 2.6–2.7 |
| granite | 2.65 | 2.6–2.7 |
| trap rock | 2.90 | 2.7–3.0 |

One sack of cement weighs 94 lb and is assumed to occupy an area of 1 ft³. One cubic foot of water contains 7.48 gal. In loose piles, most aggregate materials weigh about 100 lb per ft³.

As an example, suppose that 1 gross cubic foot of gravel weighs 106 lb per ft³, with a specific gravity of 2.65. Then the space occupied by the gravel particles is

$$\text{solid volume} = \frac{106}{2.65 \times 62.4} = 0.64 \text{ ft}^3$$

If seven sacks of cement are used per cubic yard of concrete, the absolute volume of the cement is

$$\frac{94 \times 7 \text{ sacks}}{3.15 \times 62.4} = 3.35 \text{ ft}^3$$

Suppose also that the number of gallons of water required are four per sack of cement. Then 28 gal of water are used per cubic yard of concrete. The number of cubic feet of water in 28 gal is

$$\frac{28 \text{ gal}}{7.48 \text{ gal/ft}^3} = 3.74 \text{ ft}^3$$

The total volume of cement paste is the sum of the volumes of water and cement:

$$3.35 + 3.74 = 7.09 \text{ ft}^3 \qquad \text{(this assumes no voids)}$$

To calculate the weights of ingredients to produce 1 yd³ of concrete, the following calculation is used:

$$\frac{\text{wt water}}{62.4} + \frac{\text{wt cement}}{62.4 \times \text{sp. gr.}} + \frac{\text{wt fine aggregate}}{62.4 \times \text{sp. gr.}} + \frac{\text{wt coarse aggregate}}{62.4 \times \text{sp. gr.}} = 27 \text{ ft}^3$$

**Example.** A 3200-psi concrete is to be mixed without air entrainment. As a margin of safety, a water–cement ratio of 6 gal/sack will be selected from Fig. 47. The proportions of cement to fine aggregate to coarse aggregate will be $1:2\frac{1}{2}:3\frac{1}{2}$ by weight.

The proportions by weight are

| | | | |
|---|---|---|---|
| Water | 0.53 | Sand | 2.50 |
| Cement | 1.00 | Gravel | 3.50 |

The weights per sack of cement (94 lb) are

| | | | |
|---|---|---|---|
| Water | 50 | Sand | 235 |
| Cement | 94 | Gravel | 329 |

The specific gravities are 3.15 for cement and 2.65 for the aggregates. Using the formula

$$ft^3 = \frac{wt}{62.4 \times sp.\ gr.}$$

the volumes per sack of cement are

| | | | |
|---|---|---|---|
| Water | 0.80 | Sand | 1.42 |
| Cement | 0.48 | Gravel | 1.99 |

giving a total volume of 4.69 ft³. Assume also 2 per cent voids. With voids the total volume is 4.69 × 1.02, or 4.78 ft³. So the total weight of the batch is 708 lb, and the unit weight = 708/4.78, or 148 lb per ft³.

## 5.10 slump test and ball penetration test

There is no test method available that measures directly the workability of concrete. The *slump test* is used for this purpose all over the world, although the same slump actually can be recorded for different workabilities. Despite its inadequacies, the slump test is a convenient and useful field test as a check on the variation in materials being mixed for concrete. An increase in slump may indicate that the moisture content or the grading of the aggregate has unexpectedly changed. The test is quickly performed, and since it can be executed at the concrete mixer, it can give the operator warning of an off-mix or too low or high a slump.

The standard slump cone is 12 in. high and 8 in. in diameter at the bottom (Fig. 5.12). This cone is filled from the top or small end in three equal layers, each layer being rodded 25 times with a standard $\frac{5}{8}$-in.-diameter round-nosed tamping rod. The cone is filled and struck off level with a

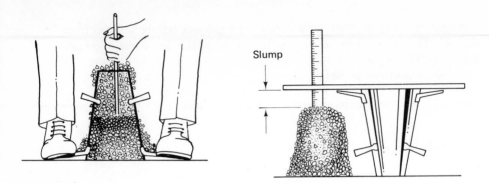

Fig. 5.12 Slump test.

trowel. It is then lifted carefully from its contents so that they are not disturbed or displaced, and the amount of slump or collapse of the cone of concrete is measured. The slump equals 12 minus the inches of height after subsidence, measured to the nearest $\frac{1}{4}$ in.

A stiff mix may have zero slump and is unlikely to give variations in slump for different workabilities. Rich mixes will be more sensitive and their slumps will indicate variations in workability. The slump test must not be delayed too long after the concrete is mixed, or the concrete will have begun to set, with a consequent reduction in slump.

The relationship of slump to workability is as follows:

| Slump (in.) | Workability |
|:-----------:|:-----------:|
| 0–1 | very low |
| 1–2 | low |
| 2–4 | medium |
| 4–7 | high |

Recommended slumps (in inches) for various types of pours are these:

1. Reinforced foundations and footings, 2–5 in.
2. Plain footings and substructures, 1–4 in.
3. Slabs, beams, and reinforced walls, 3–6 in.
4. Columns, 3–6 in.
5. Pavement, 2–3 in.
6. Massive pours, 1–3 in.

The *ball penetration test*, like the slump test, is a method of comparing one mix with another for workability. The ball test may be applied to concrete in the wheelbarrow or even in the forms after pouring. The ball apparatus consists of a cylinder with hemispherical bottom and an attached handle, the unit weighing 30 lb. The apparatus is set on a surface of the freshly poured concrete, with the handle vertical. The ball is allowed to sink into the concrete under its own weight. After it has ceased to penetrate the concrete, the ball penetration is read from the scale on the handle, to the nearest $\frac{1}{4}$ in. A minimum of three readings should be taken from each batch of concrete.

## 5.11 strength tests of concrete

The compressive strength test of cylindrical specimens is the common quality-control test or acceptance test for concrete. The strengths are usually for 7- or 28-day curing periods.

The standard size of specimen is 12 in. high and 6 in. in diameter. The mold is filled in three equal layers, each being rodded 25 times with the standard tamping rod. The full mold is struck off level with a trowel and covered with a metal or glass plate to prevent evaporation of water. The specimen is removed from the mold after 24 hr and cured for the designated number of days. Before testing, the ends of the specimen are capped with a neat cement paste or other material to provide a smooth end surface, or else ground flat, smooth, and perpendicular to the sides. The specimen is loaded in a compression-testing machine. In determining the stress, the diameter of the cylinder is measured to the nearest 0.01 in. by averaging two measurements at right angles to each other.

For a *flexural* (*bending*)-*strength test*, the concrete specimen is formed as a beam of rectangular section, with a length at least 2 in. greater than three times its depth. The smaller cross-sectional dimension must be at least three times the maximum size of the coarse aggregate used, and in any case the beam must have a cross section not smaller than $6 \times 6$ in. The mold is filled in two equal layers, with each layer rodded one stroke for each 2 in.² of area. After the curing period, the concrete beam is loaded to failure and the modulus of rupture is calculated. The loading method and other details are given in ASTM C78.

## 5.12 concrete additives

A greater weight of concrete than of any other material is used in construction. Concrete is unusual in other respects: it is a complex material that often is blended and manufactured on the jobsite. In addition, the curing of concrete is a complex chemical process that must be suitably controlled by the contractor.

Concrete does not cure properly under any and all circumstances. It does not, for example, cure at freezing temperatures. On a hot, dry, and sunny or windy day, there will be excessive evaporation of water from the exposed surface of the fresh concrete, and unless curing conditions are well controlled, a weak concrete of poor quality and surface cracks will develop. High temperatures provide too rapid a curing rate, such that there may be insufficient time to work and finish the pour. The contractor must, therefore, be able to control the setting, curing, and other conditions by various means. One of these methods of control is the use of additives to the concrete.

The specifications for the concrete may call for special characteristics, such as resistance to freezing and thawing cycles. Industrial concrete floors must be resistant to impact, wear, and abrasion. Other concretes must be impervious to water. There are additives that provide these and other characteristics. An extensive technology of concrete additives has been developed, giving the contractor greater control over curing and other characteristics. In addition to cement, water, sand, and stone, additives are now the fifth ingredient in concrete.

Additives must be used with judgment, and generally also with the advice of the supplier. Many concrete-mix designs require modification when certain admixtures are used in the concrete. Most of the organic chemical admixtures are influenced in their effect on concrete by the cement type (or even brand), the water–cement ratio, and even the temperature. Some additives have "side effects" that must be guarded against. One of the oldest additives is calcium chloride, an accelerator, which if carelessly used, may corrode reinforcing steel. The benefits of admixtures therefore are obtained only with knowledge, experience, and skill. In particular, additive materials are not curatives for inferior materials or mixtures, or for bad concrete practice.

## 5.13  air-entraining agents

Freshly mixed concrete contains about 5 per cent of air. Compaction of the concrete in the forms reduces such porosity to 1 or 2 per cent. This residual amount of air communicates with the surface of the concrete by capillary pathways that provide an entrance into the concrete for water and solutions, which may compromise the durability of the concrete. Water that enters the concrete in this way may freeze and cause spalling.

It is standard practice to add an *air-entraining agent* to concretes subject to freezing and thawing. The air-entraining agent increases the porosity of the concrete to 4–6 per cent, but has the important effect of breaking up the network type of porosity into disconnected bubbles of air. These artificially introduced voids have a diameter of about 0.002 in., and there are about $\frac{1}{2}$ billion of them in 1 yd³ of air-entrained concrete.

Figure 5.10 indicates that air entrainment will result in a 20 per cent loss of strength. This is compensated for by reducing the water–cement ratio. But this may reduce the slump, making for difficulty in placing and working the concrete. But the entrained air bubbles function as additional fine aggregate, lubricating the mix and making possible a reduction in mixing water of as much as 15 per cent. The amount of air-entraining agent must be adapted to the mix, since less entrained air is used for very coarse aggregate.

The success of air entrainment against freezing is explained by the fact that the multitude of air bubbles provide space to absorb the volume expansion when water freezes.

The air-entraining agents are chemical surfactants (surface-active agents) which reduce the surface tension of water. Such a chemical enables water to hold air; the water thus becomes a foam. Not all surfactants, however, are compatible with concrete and with concrete reinforcing steel. Suitable agents include lignins extracted from wood, sulfonated hydrocarbons, certain fats and oils, and, more recently, polymers. The latter, which are chemicals of high molecular weight, are discussed in Chapter 14.

## 5.14 accelerators

An additive that will advance the initial set of a concrete is called an *accelerator*. Accelerators may be required for a variety of reasons. The curing period may have to be reduced for earlier removal of forms, or to reduce the time for a concrete repair job. Accelerators are often used in cold weather to shorten the period during which protection is needed for the concrete.

Calcium chloride in amounts of about 2 per cent is commonly used for this purpose. Two per cent is approximately 2 lb per bag of cement. Greater amounts than this are not used. The use of an accelerator will increase the rate of heat generated by setting. This is an aid in preventing freezing of the concrete. Adverse effects of calcium chloride include a somewhat increased shrinkage and a lessened resistance to sulfate attack.

Other accelerators are chiefly proprietary chemicals sold under brand names. Some of these compounds are not suitable for prestressed concrete.

## 5.15 retarders

The purpose of a *retarder* is to delay the setting time of the concrete. In hot weather and especially with high winds and sunlight, hydration is accelerated, and in such conditions there may be insufficient time to place, consolidate, screed, and finish the concrete. For example, ready-mix trucks may have to stand in the hot sun for excessive periods of time if there are delays at the jobsite. The rapid evaporation of water from the surface of cement, together with rapid setting and high heat generation, may result in a network of surface cracks.

Retarders control the otherwise rapid hydration, retaining the water for workability. Most retarders allow a reduction in the water required without reducing the slump. The reduction in water may be an important consideration for thin-shell concrete in barrel roofs, domes, and similar shapes. Such concrete shells require a mix of high workability and cohesiveness, but with a low slump, to prevent the concrete from sagging on the steep slopes of the roof. Retarders are used in greater amounts for higher temperatures but generally should not be used for temperatures below 60°F.

Retarders may be cellulose chemicals, such as sucrose (table sugar), calcium lignosulfonate, or organic acids, or their salts, such as citric or tartaric acid, proteins, phosphates, or borates. A 0.05 per cent admixture of starch or sugar will retard the initial setting of portland cement by about 4 hr. Workability and 28-day strength will also be improved. Increased amounts of these retarders will further delay initial hardening.

## 5.16 dispersants

Water does not readily wet cement. The cement particles tend to clump together in the presence of water, and when this occurs water cannot reach all the particles to hydrate them.

Certain *dispersants* such as calcium lignosulfonate will separate such particles by giving them an electrostatic charge to repel them from other particles equally charged. The dispersing agent will also improve the workability, since a film of water lubricates all the particles of cement. Most of the commercial dispersants also contain air-entraining agents. If used in excessive amounts they will usually entrain excessive air. They are very effective in increasing the strength of pozzolan concretes.

## 5.17 hardeners

Certain concrete surfaces, chiefly floors, must be resistant to impact, abrasion, and wear, and must also be nondusting under conditions of traffic flow. Two types of *hardeners* are used to meet these service conditions. One of the types is a fine metallic or mineral aggregate, which is added into the surface of the freshly poured concrete. The other is a chemical hardener.

The metallic or mineral hardeners are graded iron, quartz, flint, emery, or other particles or aggregates. These are mixed dry with portland cement. They are spread evenly over the surface of the fresh concrete and worked into the surface.

The chemical types of hardeners are liquid solutions of fluosilicates which combine with free lime and calcium carbonate in the concrete to produce a very hard, nondusting surface that is resistant to wear. Polyethylene containers should be used for holding these chemicals, since they attack metals.

### 5.18 other additives

Colored concrete is often required, especially for floors. Suitable paints for concrete surface are available, but even the specially formulated traffic paints have a limited life. For permanent color, the concrete surface must be pigmented. Stable colors suitable for concrete pigmenting are largely restricted to browns, green, blue, gray, and beige, all made from metallic oxides that are fast colors in sunlight. Such colors are usually blended into a topping mix applied to the concrete, or shaken dry over the fresh concrete and then troweled or floated in.

Water repellents are used to prevent water permeation through concrete. Surface sealing additives seal the pores in the surface of the hardened concrete to prevent permeation of water and the absorption of oils or other spilled materials.

Certain finely divided materials serve as improvers to the workability of concrete. Such materials may be bentonite clay, kaolin, chalk, or similar mineral materials.

Fig. 5.13 Grouting applications.

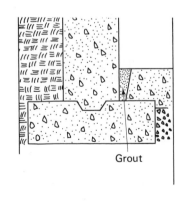

Grout

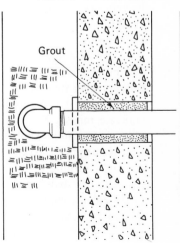

Grout

Some architectural surfaces require an exposed aggregate in order to obtain an attractive appearance. To prevent the cement paste from bonding to aggregate that requires exposure, set-inhibiting agents are applied to the exposed area, either as a liquid applied to the forms of vertical surfaces, or as a powder applied to a fresh surface. The depth to which such an inhibitor penetrates increases with the quantity used. After the concrete has sufficiently hardened, the retarded cement is brushed off or hosed to expose the aggregate.

*Grouts* are nonshrinking concrete filler and jointing materials. Two examples of grouting are shown in Fig. 5.13. In the first example the grout is used as a foundation seal to counteract the shrinkage of the standard concrete members. In the second example the grout is used to caulk a pipe opening in the concrete. If an ordinary concrete were used in these applications, the mixing water would have to be reduced to minimize shrinkage of the concrete, but this would make the mix difficult to place. In any case, ordinary concrete may shrink sufficiently to pull away from the concrete against which it is poured.

Anchor bolts and machine bases are other common applications for the use of expanding grouts. The expansion in such grouts is obtained either by polymer additives or by iron. The iron oxidizes to iron oxide, which occupies a larger volume than the original iron.

## 5.19   storage and handling

Because the ever-present water vapor of the atmosphere can harden cement, given sufficient time, cement has only a limited "shelf life." Cement that is 4 months old should be tested before being used. At all times cement must be kept dry. Cement in multiwall paper bags must be stored in weatherproof conditions. Storage buildings for such bagged cement should have a floor raised above ground and protected by a vapor barrier such as polyethylene. Bags are best stored on pallets so that they are elevated above the floor. The building roof should be pitched so that condensation will run along the ceiling rather than drip.

Hardened cements should be removed from the inside surfaces of concrete mixers. During lunch breaks or other prolonged delays, the concrete mixer can be kept running with some water and a couple of shovelfuls of stone or gravel. At the end of the day the mixer must be cleaned by hosing out and churning with water and stone.

The first batch put into a clean and dry mixer should contain a little extra cement, sand, and water so that the interior of the mixer can be coated without upsetting the intended concrete mix. The mixing time for small mixers with a capacity of 1 yd³ or less should be at least 1 minute. For larger capacities the minimum mixing time should be increased $\frac{1}{4}$ minute for each yard additional (or fractional yard).

The grinding action of prolonged mixer operation will increase the proportion of fines in the concrete mix. This effect results in increased surface area and water absorption of the aggregate. The end result is a change in slump.

If a mixture of fine and coarse solid materials is dumped in a pile, the larger particles tend to roll down the pile to the bottom while the finer fractions remain in place. The dumping operation tends to separate fine and coarse fractions. This separation of size fractions is called *segregation*. Handling and placing operations on concrete must be so controlled that segregation does not occur, since it is virtually impossible to correct this condition. Segregation is promoted by making concrete flow horizontally or on slopes. Concrete must be placed in horizontal layers about 2 ft deep, each layer being compacted with vibrators or puddled with sticks and shovels, and spaded at the forms so that voids will be closed at the surface of the concrete.

## PERTINENT ASTM SPECIFICATIONS

C150-68 (Part 9) Standard Specification for Portland Cement

C187-68 (Part 9) Test for Normal Consistency of Hydraulic Cement (Vicat Apparatus)

C191-65 (Part 9) Time of Setting of Hydraulic Cement by Vicat Needle

C31-69 (Part 10) Making and Curing Concrete Compression and Flexural Strength Test Specimens

C33-67 (Part 10) Concrete Aggregates

C39-66 (Part 10) Compressive Strength of Molded Concrete Cylinders

C40-66 (Part 10) Organic Impurities in Sands for Concrete

C70-66 (Part 10) Surface Moisture in Fine Aggregate

C78-64 (Part 10) Flexural Strength of Concrete

C94-69 (Part 10) Specification for Ready-Mixed Concrete

C127-68 (Part 10) Specific Gravity and Absorption of Coarse Aggregate

C128-68 (Part 10) Specific Gravity and Absorption of Fine Aggregate

C136-67 (Part 10) Sieve Analysis of Fine and Coarse Aggregate

C138-63 (Part 10) Weight per Cubic Foot, Yield, and Air Content of Concrete

C143-69 (Part 10) Slump of Portland Cement Concrete

C150-69 (Part 10) Standard Specification for Portland Cement

C172-68 (Part 10) Sampling Fresh Concrete

C173-68 (Part 10) Air Content of Freshly Mixed Concrete by the Volumetric Method

C192-69 (Part 10) Making and Curing Concrete Test Specimens

C360-63 (Part 10) Ball Penetration Test for Fresh Concrete
C294-69 (Part 10) Standard Descriptive Nomenclature of Constituents
of Natural Mineral Aggregates
C617-68 (Part 10) Capping Cylindrical Concrete Specimens

## QUESTIONS

1 How does the relative amount of mixing water added to concrete affect the drying shrinkage of the concrete?

2 (a) What damage is done by excessive magnesia in cement?
(b) Does an excess of free lime have the same effect?

3 What is the difference between cement clinker and portland cement?

4 What purpose does gypsum serve in cement?

5 What would the effect be in cement of (a) an excess of gypsum, (b) too little gypsum?

6 Compare $C_2S$, $C_3S$, and $C_3A$ as to heat released during setting.

7 What is meant by hydration of cement?

8 What is a hydraulic cement?

9 What reaction is first to occur when cement sets?

10 What advantages are offered by steam curing concrete?

11 Indicate the types of application for high early strength, low-heat, lumnite, and sulfate-resistant cements.

12 What is a pozzolan?

13 Can a pozzolan alone be a cementing material?

14 Why are flat shales not desirable for concrete aggregate?

15 What is meant by a water–cement ratio?

16 What advantages do air-entraining agents provide in a concrete?

17 What circumstances would require (a) an accelerator; (b) a retarder in concrete?

18 What is a grout?

19 What influence do the four basic chemical constituents of portland cement exert on its characteristics?

20 For what purpose is calcium chloride added to cement?

21 What determines the compressive strength of a concrete?

22 Describe an air-entrained concrete.

23 What are some of the methods by which shrinkage of concrete may be reduced?

24 Concrete can be made with cement and sand. What are the reasons for using coarse aggregate?

**25** Explain how the quality and characteristics of a concrete may be altered during (a) mixing; (b) delivery by ready-mix truck; (c) placing in the forms.

**26** What methods and materials can be used to assist the hardening of concrete poured in cold weather?

**27** How are the following characteristics obtained in concrete: (a) high strength; (b) high durability; (c) heat resistance; (d) low shrinkage; (e) low thermal conductivity; (f) impermeability.

**28** If a sand of specific gravity 2.65 weighs 105 lb per ft$^3$ as piled, what percentage of the volume of the pile is voids?

**29** What is the effect of air entrainment on workability?

**30** (a) What is the effect of a dispersant on cement?
(b) How does a dispersant influence workability?

**31** A concrete is to be formulated with a mix of 1 cement:1.75 sand:3.5 coarse aggregate by volume, with a water–cement ratio of 0.55 by weight. The sand weighs 95 lb per ft$^3$ and the coarse aggregate 88 lb per ft$^3$. Determine the absolute volume of each component in cubic feet per bag of cement, the number of bags of cement per cubic yard, the dry sand and coarse aggregate per cubic yard, and the water per cubic yard. Use standard specific gravities of 3.15 for cement and 2.65 for aggregates. Assume no voids.

**32** A concrete is to use $6\frac{1}{2}$ bags cement per cubic yard and 6 gal of water per bag. The volume of fine aggregate will be half the volume of coarse aggregate. Specific gravities are as given in Question 31; the loose fine aggregate weighs 95 lb per ft$^3$ and the coarse aggregate 88 lb per ft$^3$. Assume no voids. Find the absolute volumes of cement paste and total aggregate per cubic yard.

**33** A heavy-duty concrete floor is to use a water–cement ratio of 0.4 with 10 bags of cement per cubic yard. Cement–sand–stone proportions by volume are 1 : 1 : 2. Using the standard data of the previous question, determine the absolute volumes of cement paste and fine and coarse aggregate. Assume no voids.

Concrete structural products and masonry block are factory produced. The better and more consistent quality control obtainable under factory conditions result in dependable high quality, which is not always obtained in on-site mixing and curing of concrete.

## 6.1 precast structural concrete

## members

Precast structural concrete includes floor slabs, roof slabs, wall panels, columns, beams and joists, piles, and specialties such as stairs. High-strength concrete is usual in these products, which generally require reinforcement with steel bar or wire. The reinforced members may be normally reinforced, such that the only stress in the steel reinforcing is that imposed by the building loads, or prestressed. Prestressed reinforcing is stressed during manufacture of the concrete member, the prestressing being of such a nature that it assists the concrete member in resisting imposed building loads.

# PRECAST
# CONCRETE
# PRODUCTS

# 6

Typical prestressed concrete products include

1. bridge girders
2. complete frames for prefabricated buildings
3. large concrete pipe
4. floor beams and joists
5. floor and road slabs
6. lighting standards
7. bearing piles
8. roof beams and purlins

When these units are produced in quantity, their cost is comparable with the cost of ordinary reinforced concrete for the same purpose. The prestressed members, however, are lighter in weight and section. Probably the outstanding advantage of the prestressed products is the avoidance of formwork at the jobsite.

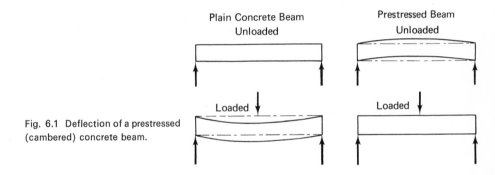

Fig. 6.1 Deflection of a prestressed (cambered) concrete beam.

Prestressed concrete may be pretensioned or posttensioned. The pretensioning method stresses the reinforcing strands in the forms before the concrete is poured around them. The concrete sets and bonds to the steel. When the prestressing jacks are released, stress is transferred by the steel to the concrete. The pretensioned member then has a slight camber as a result of the artificially induced compression in the concrete. The prestressed side of the concrete beam thus has an increased ability to sustain a tension force (Fig. 6.1). The initial prestress in the steel reinforcing is greater than the intended stress in the steel, owing to a loss of prestress occasioned by the following factors:

1. Elastic deformation of the concrete when load is transferred to it.
2. Shrinkage of the concrete as a result of long-term hardening.
3. Creep in the concrete.
4. Creep in the steel.

In posttensioning the reinforced structural member is poured and cured, leaving a number of passages through which the posttensioning cables are passed. After hardening of the concrete, the reinforcing is tensioned by hydraulic jacks.

High-strength steel is used in prestressing, either as alloy rod $\frac{7}{8}$ to $1\frac{3}{8}$ inches in diameter or as stranded cable, usually 7 or 19 wires per strand. Ultimate tensile strength for such reinforcing lies in the range 190,000–240,000 psi, and the final stress after the loss of prestress mentioned above ranges from 115,000 to 132,000 psi. Most structural steels have ultimate tensile strengths in the range of about 70,000 psi. These high-strength wires have only limited elongation, about 4 per cent in 24 in. Therefore, the load, and the strain, on prestressing wires must be closely controlled. Usually the force is determined to within about $2\frac{1}{2}$ per cent accuracy, using various types of gages or a dynamometer.

## 6.2 modular concrete block

Concrete block is made in both standard and lightweight concrete. A wide range of sizes and types is manufactured for interior and exterior use, including such specialties as chimney block.

Fig. 6.2 Basic concrete block sizing. This basic size of block is related to a standard brick (Fig. 5.13) in that it is two bricks wide, three brick courses high, and two bricks long.

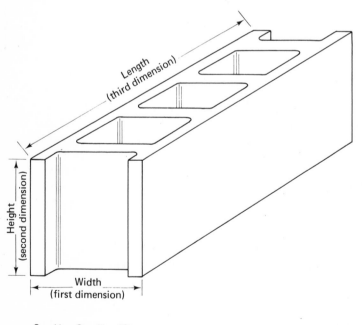

8 X 8 X 16
Width X Height X Length

Fig. 6.3 Some types and sizes of concrete block.

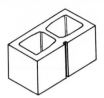

Center Scored

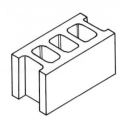

Stretcher (3 core)

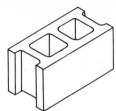

Stretcher (2 core)

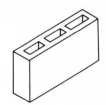

Partition

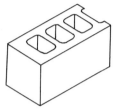

Corner

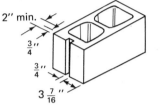

2" min.
$\frac{3}{4}$"
$\frac{3}{4}$"
$3\frac{7}{16}$"

8 × 8 × 16
Sash

$7\frac{5}{8}$"

8 × 8 × 8
Half-Sash

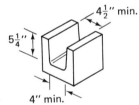

$4\frac{1}{2}$" min.
$5\frac{1}{4}$"
4" min.

8 × 8 × 8
Standard Lintel

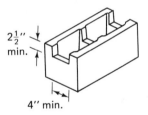

$2\frac{1}{2}$" min.
4" min.

8 × 8 × 16
Bond Beam

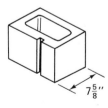

$7\frac{5}{8}$"

12 × 8 × 8
Half-Sash

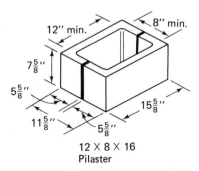

12" min.
8" min.
$7\frac{5}{8}$"
$5\frac{5}{8}$"
$15\frac{5}{8}$"
$11\frac{5}{8}$"
$5\frac{5}{8}$"

12 × 8 × 16
Pilaster

*precast concrete products* / **105**

Fig. 6.3 (cont.)

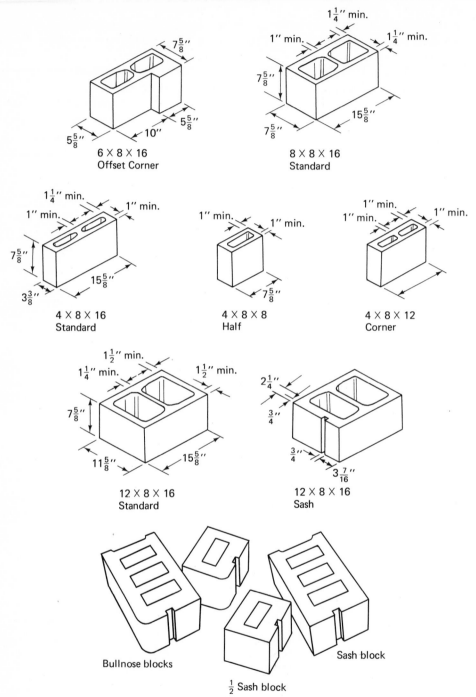

6 × 8 × 16
Offset Corner

8 × 8 × 16
Standard

4 × 8 × 16
Standard

4 × 8 × 8
Half

4 × 8 × 12
Corner

12 × 8 × 16
Standard

12 × 8 × 16
Sash

Bullnose blocks

½ Sash block

Sash block

Modular size is usually given by the nominal or modular dimension, as with lumber. Thus an $8 \times 8 \times 16$ in. block measures $7\frac{5}{8} \times 7\frac{5}{8} \times 15\frac{5}{8}$ in. In giving the block size, the first number is the width, the second is the height, and the third is the length of the block (Fig. 6.2). The difference between nominal and actual size of $\frac{3}{8}$ in. is accounted for by a $\frac{3}{8}$-in. mortar joint, this combination giving a 4-in. standard module. The actual thickness of the mortar joint as laid is adjusted so that the length of the wall may be made up of an integral number of whole blocks, but must lie between $\frac{5}{16}$ and $\frac{1}{2}$ in. The modular design of these blocks will be apparent in an examination of the block sizes of Fig. 6.3, which shows a selection of the many sizes and shapes available. Many types of block are also made in half-size dimensions. The sash block shown in the figure is required for anchoring window and door frames, the method being illustrated in Fig. 6.4.

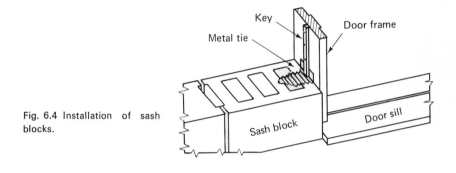

Fig. 6.4 Installation of sash blocks.

Five basic types of standard concrete masonry units are produced:

1. a. hollow load-bearing
   b. solid load-bearing
2. hollow non-load-bearing
3. building tile
4. concrete brick
5. screen blocks

Screen blocks (Fig. 6.5) are manufactured in various designs. These units are used to shield out street noises and to provide some degree of privacy or removal, but to allow entry of daylight.

A solid concrete block is defined by the ASTM as one that has a core area not exceeding 25 per cent of the total cross-sectional area. A hollow block is one in which the core area (hollow area) exceeds 25 per cent; for such a block the core area is usually in the range 40–50 per cent of gross area.

Concrete block shapes are produced in block-forming machines by pressure and vibration, then steam-cured. They are produced to specification for compressive strength, moisture content, absorption, and other charac-

Fig. 6.5 Screen blocks used in front of and behind the glass panels of an office building facing west.

teristics. Absorption is a measure of the density of the concrete in the block. Concretes shrink with loss of moisture, and this characteristic requires that block meet a moisture-absorption specification. If moist block were built into a wall, the wall construction would restrain the block against its own shrinkage, and the result would be cracked block. Moisture absorption and the several grades of concrete masonry units are defined in ASTM C90-66T.

## 6.3 allowable loads for

## concrete block

The compressive strength of the block determines its allowable compressive or bearing load. This allowable stress is dictated by building codes. The specified minimum compressive strength for block is given as pounds per square inch averaged over the gross area of the block, including core area. ASTM specifications require the following minimum strengths and water absorption for concrete block:

|  | Minimum Face-Shell Thickness, (in.) | Compressive Strength, Minimum, psi, Average Gross Area | | Water Absorption, Maximum, lb per ft³ of Concrete, Average of 5 Units |
|  |  | Average of 5 Units | Individual Units |  |
| --- | --- | --- | --- | --- |
|  | $1\frac{1}{4}$ or over : |  |  |  |
| hollow load-bearing | grade A | 1000 | 800 | 15 |
| concrete masonry | grade B | 700 | 600 | — |
| units, ASTM C90 | under $1\frac{1}{4}$ and over $\frac{3}{4}$ | 1000 | 800 | 15 |
| hollow non-load-bearing concrete masonry units, ASTM C129 | not less than $\frac{1}{2}$ | 350 | 300 |  |
| solid load-bearing concrete masonry units, ASTM C145 : |  |  |  |  |
| Grade A |  | 1800 | 1600 | 15 |
| Grade B |  | 1200 | 1000 | 15 |

Grade A blocks are used above and below grade where exposed to frost action; grade B blocks are used in locations where they are protected from frost action.

A factor of safety of 4 or better is used for block construction. The allowable load of a block wall is influenced by the pattern of mortar bedding, which may be the full mortar bed of Fig. 6.6(a) or the face-shell bedding of (b). A wall laid with full bedding has a bearing strength of 53 per cent of the block; with face-shell bedding the bearing strength of the wall is reduced to 42 per cent of the block. A full mortar bed is preferred. Finally, allowable wall compressive stress is influenced by the type of mortar. For cement mortar the maximum allowable compressive stress is 100 psi in load-bearing walls, 80 psi for cement-lime mortar.

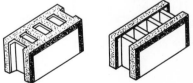

Fig. 6.6 Full mortar bed and face-shell mortar bed.

The cement or cement-lime mortar allowed in block construction must use a proper masonry cement with good workability and retentivity for water. Retentivity is important, since water may be lost to the concrete masonry units, which often have a high absorption. Properly graded sand and a proper masonry cement assist in retaining water. The following are suitable mortar mixes by volume:

| Service | Cement | Hydrated Lime or Lime Putty | Mortar Sand, Damp and Loose |
|---|---|---|---|
| ordinary | 1 masonry cement or | — | 3 |
|  | 1 portland cement | $1-1\frac{1}{4}$ | 4–6 |
| severe frost action | 1 masonry cement plus 1 portland cement | — | 4 |
|  | 1 portland cement | $\frac{1}{3}$ | $2\frac{1}{2}$–3 |

Fig. 6.7 Bond patterns for concrete block.

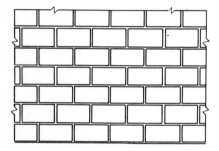

Common or Running Bond

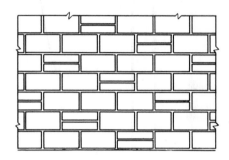

Coursed Ashlar

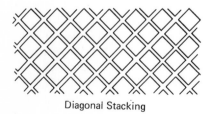

Stack Bond

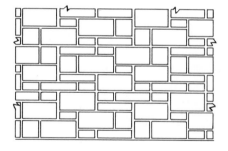

Random Ashlar

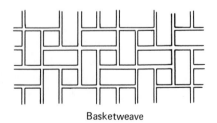

Diagonal Stacking

Basketweave

The wide range of shapes and sizes of concrete block allows a range of wall patterns to be used. Some of these are shown in Fig. 6.7. The standard *running bond* or *half-bond*, usual in brickwork, is commonly used. The *stack bond* is a combination of vertical and horizontal courses. *Diagonal stacking* is perhaps less common and is usually restricted to a square block 8 × 8. *Basketweave* and various ashlar patterns are also used to offer an interesting pattern. Sometimes some of the block are laid in relief. Block with a patterned face and screen blocks offer other possibilities for patterns.

    *Lintel blocks* (Fig. 6.8) are filled with reinforced concrete to provide a lintel over doors and windows. Lintel blocks are used in the same way to provide a solid reinforced concrete beam in a block wall to reinforce the block construction.

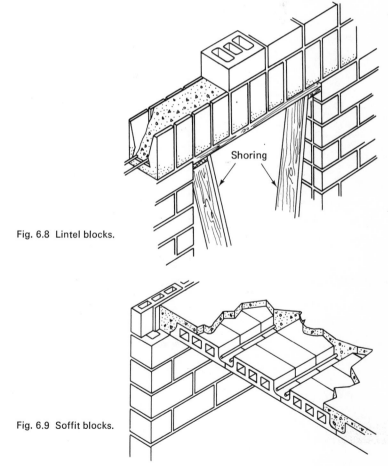

Fig. 6.8 Lintel blocks.

Fig. 6.9 Soffit blocks.

# U FACTOR FOR BLOCK WALLS

| Basic Wall Construction | Interior Finish | | | |
| --- | --- | --- | --- | --- |
| | Plain Wall, No Plaster | Wall Direct | ½-in. Plaster on: ¼-in. Furring with: | |
| | | | ⅜-in. Plaster Board | ½-in. Rigid Insulation |
| concrete — 8-in. sand and gravel or limestone | 0.53 | 0.49 | 0.31 | 0.22 |
| masonry (cores not filled) — 8-in. expanded slag, clay, or shale | 0.33 | 0.32 | 0.23 | 0.18 |
| 12-in. sand and gravel or limestone | 0.49 | 0.45 | 0.30 | 0.22 |
| 12-in. expanded slag, clay, or shale | 0.32 | 0.31 | 0.23 | 0.18 |
| concrete — 8-in. sand and gravel or limestone | 0.39 | 0.37 | 0.26 | 0.19 |
| masonry (cores filled with insulation) — 8-in. expanded slag, clay, or shale | 0.17 | 0.17 | 0.14 | 0.12 |
| 12-in. sand and gravel or limestone | 0.34 | 0.32 | 0.24 | 0.18 |
| 12-in. expanded slag, clay, or shale | 0.15 | 0.14 | 0.12 | 0.11 |
| cavity walls (with 2-in. or larger cavity; cavity not filled with insulation) — 10-in. wall of two 4-in. sand and gravel or limestone units | 0.34 | 0.33 | 0.24 | 0.18 |
| 10-in. wall of two 4-in. cinder, expanded-slag, clay, or shale units | 0.26 | 0.24 | 0.19 | 0.15 |
| 10-in. wall of 4-in. face brick and 4-in. sand and gravel or limestone units | 0.38 | 0.36 | 0.25 | 0.19 |
| 10-in. wall of 4-in. face brick and 4-in. cinder, expanded-slag, clay, or shale unit | 0.33 | 0.31 | 0.23 | 0.18 |
| 14-in. wall of 4-in. face brick and 8-in. sand and gravel or limestone units | 0.33 | 0.31 | 0.23 | 0.18 |

| | | | | |
|---|---|---|---|---|
| 14-in. wall of 4-in. face and brick and 8-in. cinder, expanded-slag, clay, or shale unit | 0.26 | 0.25 | 0.19 | 0.16 |
| 14-in. wall of 4-in. and 8-in. sand and gravel or limestone units | 0.30 | 0.28 | 0.21 | 0.17 |
| 14-in. wall of 4-in. and 8-in. cinder, expanded-slag, clay, or shale units | 0.22 | 0.21 | 0.17 | 0.14 |
| 4-in. face brick plus: | | | | |
| 4-in. sand and gravel or limestone unit | 0.53 | 0.49 | 0.31 | 0.23 |
| 4-in. cinder, expanded-slag, clay, or shale unit | 0.44 | 0.42 | 0.28 | 0.21 |
| 4-in. common brick | 0.50 | 0.46 | 0.30 | 0.22 |
| 8-in. sand and gravel or limestone unit | 0.44 | 0.41 | 0.28 | 0.21 |
| 8-in. cinder, expanded-slag, clay, or shale unit | 0.31 | 0.30 | 0.22 | 0.17 |
| 8-in. common brick | 0.36 | 0.34 | 0.24 | 0.19 |
| 1-in. wood sheathing, paper, 2 × 4 studs, wood lath and plaster | — | 0.27 | 0.27 | 0.20 |
| 4-in. common brick plus: | | | | |
| 4-in. sand and gravel or limestone unit | 0.45 | 0.42 | 0.28 | 0.21 |
| 4-in. cinder, expanded-slag, clay, or shale unit | 0.38 | 0.36 | 0.26 | 0.19 |
| 8-in. sand and gravel or limestone unit | 0.37 | 0.35 | 0.25 | 0.19 |
| 8-in. cinder, expanded-slag, clay, or shale unit | 0.23 | 0.27 | 0.20 | 0.16 |
| 8-in. common brick | 0.31 | 0.30 | 0.22 | 0.17 |
| 1-in. wood sheathing, paper, 2 × 4 studs, wood lath and plaster | — | 0.25 | 0.25 | 0.19 |
| wood frame   wood siding, 1-in. wood sheathing, 2 × 4 studs, wood lath and plaster | — | 0.25 | 0.24 | 0.19 |

Source: Portland Cement Association.

*Soffit blocks* are used in floor construction together with poured concrete and reinforcing, as shown in Fig. 6.9.

A *pilaster* is a built-in column in a wall. Pilaster block is used to form the two types of pilaster shown in Fig. 6.10: flush pilaster and full pilaster. Such pilaster blocks offer a larger bearing area for the support of beams carried by the wall. The pilaster block has a large core area for a poured concrete fill.

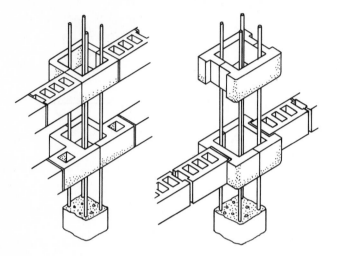

Fig. 6.10 Flush pilaster and full pilaster concrete blocks.

*Bullnose* and L-*type corner blocks* are used at the corners of the block construction. Sill blocks form a window sill in a block wall. For capping a wall or roof parapet, coping blocks with a sloping top surface are used.

## 6.5 other concrete block

### products

Lightweight block made with haydite is also produced with a glazed surface for exterior or interior finish. The glazed face may be a coat of silica sand, pigment, and cement. More recently the glazed surface has been produced with a plastic resin that cures hard and is mixed with pigment and ground stone.

Woodchip block is a lightweight block made of treated wood shavings or sawdust and cement paste. These are made for several purposes: as a stacking block, an insulating slab, or a ceiling block. The form blocks used for walls are simply stacked without mortar and filled with concrete, the blocks thus serve as a concrete form. Ceiling blocks are used to form a ribbed concrete floor in the same manner as soffit blocks discussed above. Woodchip blocks will hold nails and can be cut with woodworking tools.

Interlocking concrete blocks stacked without mortar or concrete fill and self-supporting have been used in some types of construction (Lok-a-Blok).

The cellular concrete block is not made of lightweight aggregate. This block has for its raw materials sand, lime, and aluminum powder, with water added to make a slurry of the mix. Hydrogen gas generated by a chemical reaction in the mix expands the slurry, which then sets as a cellular concrete. Curing is completed with steam in an autoclave.

Cellular concrete weighs about one-fourth as much as normal concrete. In addition to light weight, useful characteristics include good thermal and acoustic insulation and fire resistance.

Cement or cement-lime mortar is used to lay up cellular concrete block. Like woodchip blocks, cellular concrete can hold nails and is easy to cut. Such blocks are used for light construction, such as for bearing walls, roof slabs, and fireproofing of steel beams.

### 6.6   properties of concrete

### masonry units

The accompanying table gives coefficients of heat transmission ($U$ factors) for various types of construction using concrete masonry units. The hollow types of block made of lightweight aggregates are very effective in the reduction of sound transmission.

*PERTINENT ASTM SPECIFICATIONS*

C145-66T (Part 12) Solid Load Bearing Concrete Masonry Units
C90-66T (Part 12) Hollow Load Bearing Concrete Masonry Units
C140-65T (Part 12) Sampling and Testing Concrete Masonry Units
C331-64T (Part 12) Lightweight Aggregates for Concrete Masonry Units
C330-69 (Part 10) Lightweight Aggregates for Structural Concrete
C331-69 (Part 10) Lightweight Aggregates for Concrete Masonry Units

### QUESTIONS

1   When casting concrete blocks, why is the mold vibrated?

2   Why are concrete blocks steam-cured?

3  Describe the difference between face-shell and full mortar bedding.

4  (a) How is a lintel formed using precast masonry?
   (b) What are some alternative methods used to span openings?

5  Differentiate between flush-wall and full pilaster block.

6  What advantages do lightweight block offer?

7  What is the difference in application between grade A and B concrete block?

*Bricks* are solid masonry units usually made of burned clay. *Clay tiles* are hollow units of burned clay (certain clay products called "tile" are not hollow).

Two broad classes of brick may be distinguished:

1. Those made of clay and fired to obtain hardness.
2. Those made of cementing materials that harden by a chemical reaction. This group includes both concrete brick and sand-lime brick.

*Sand-lime brick* is made from a mixture of 5–10 per cent hydrated lime and silica sand, cured with steam in autoclaves. This is a strong type of brick with good resistance to frost action and to fire.

## 7.1 manufacture of brick and tile

Brick and tile are manufactured from clays and shales. While virtually any clay can be made into brick, the quality of brick would be unacceptable or

# BRICK AND TILE

# 7

would not be adaptable to modern brick manufacturing methods if the characteristics of the clay were uncontrolled. Sun-dried brick made of any available clay are called adobe, and have been used traditionally in dry geographical areas where wood was scarce.

A suitable brick clay must have the following characteristics:

1. It must be plastic when tempered with water so that it can be shaped.
2. It must be strong enough to retain its shape when "green," that is, before burning.
3. It must not warp or shrink excessively when dried.
4. It must not warp or shrink excessively when subjected to the brick-burning process.

Brick clays are mined by the open-pit process, dried, crushed, and blended. The manufacturing process begins with tempering, which is a mixing of the clay with the required quantity of water in a pug mill. The amount of water required is governed by the method of shaping the brick. There are three methods of forming the brick:

1. The stiff mud process, 12–15 per cent water by weight.
2. The dry press process, less than 10 per cent water.
3. The soft mud process, 20–30 per cent water.

The stiff mud process extrudes the clay through a die in a continuous cross section in the manner of toothpaste from a tube. The clay is de-aired in a vacuum to remove air. A wire cutter cuts the continuous shape into individual bricks. This extrusion method smears the surface of the brick, making it somewhat less porous than brick produced by the other methods. Hollow clay tile shapes must be made by the stiff mud process, since hollow shapes cannot be formed by pressing.

The soft mud and the dry press methods shape the brick by pressure in molds, very high pressures being required for the dry press method.

After forming, the green brick contain too much moisture, and if they were immediately burned, they would warp, shrink, and crack, and the excessive moisture could turn to steam and explode the brick. The next step in manufacture is to pass the brick through a drying kiln maintained at a relatively low temperature so that the moisture in the brick is reduced at a controlled rate.

The final operation is burning. This is a controlled heating and cooling process that takes the brick to its vitrification temperature, which may be 1800°F or higher. At vitrification the low-melting components of the brick liquefy to a glass, bond the brick particles, and to a limited extent fill the pores of the brick. Vitrification is a rather complex softening process, but is not melting, because the components of the brick material do not have melting points, only softening points.

Fig. 7.1 Periodic kiln for burning bruck and tile. Brickmasons are installing the floor brick, which in this type of kiln are the flues connected to the chimney. The round turrets in the wall contain the burners. The dome roof uses a brown brick at the walls and a white brick at the center of the dome.

Two general types of kilns are used for burning: the *continuous* or *tunnel kiln* and the *periodic kiln*. The tunnel kiln is a very long rectangular furnace, as much as 600 ft long in the case of a high vitrification temperature, through which the brick are slowly moved on kiln cars carried on a pair of railroad rails. A periodic kiln is shown in Fig. 7.1. This type of kiln is loaded with brick or tile, then fired, then cooled so that the brick can be withdrawn and another load of brick moved in. In this particular periodic-kiln design, the round turrets projecting from the wall of the circular kiln contain the gas burners. The passages in the hollow floor tile are exhaust passages for the flue gases. Because the flue gases go down to the floor instead of upward into a chimney, this type of kiln is also called a *downdraft kiln*. The kiln itself is one of the more interesting types of construction projects. The dome roof, 40 ft in diameter, contains 21,000 brick, laid in 101 courses without any supporting formwork during installation.

## 7.2 types of brick

As is the case with concrete masonry units, the nominal dimension of a brick is the actual dimension plus the thickness of the mortar joint (except for

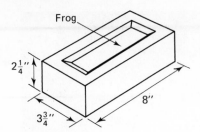

(a) Standard Dry Press Face Brick

(b) Standard Extruded Brick with Jamb Slot

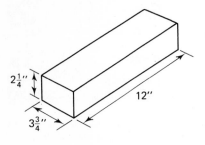

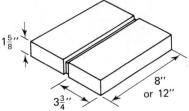

(c) Norman Brick

(d) Roman Brick

Fig. 7.2 Standard brick shapes: (a) standard dry press face brick; (b) standard extruded brick with jamb slot; (c) norman brick; (d) roman brick.

Fig. 7.3 Surface textures for face brick.

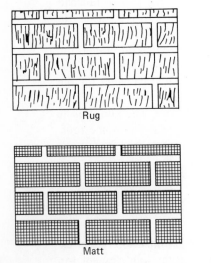

Rug

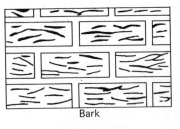

Bark

Matt

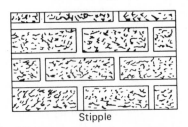

Stipple

firebrick; firebrick dimensions are actual dimensions). Generally, the thickness of a mortar joint is $\frac{1}{2}$ in. for common brick, $\frac{3}{8}$ or $\frac{1}{2}$ in. for face brick, and $\frac{1}{4}$ in. for glazed brick. Face brick are generally used for exposed walls and common brick for backup applications.

Some of the more common brick shapes are shown in Fig. 7.2. The *standard brick* of Fig. 7.2(a) contains a *frog* or depression in one face; such a brick cannot be extruded in the stiff mud process but is produced by dry pressing. Three vertical courses of this brick will give a height of 8 in. The *perforated brick* of Fig. 7.2(b) is a stiff mud brick. This brick has 10 holes and a notch $\frac{3}{4} \times \frac{3}{4}$ in. at one end for a window-jam slot. *Norman brick* are 12 in. long. *Roman brick* are produced as double brick and broken into two bricks on the jobsite. The fracture produces a rough-textured effect on the exposed edge. Other sizes of brick are in use; virtually all are either 8 or 12 in. long and give a 4-in. wall module on thickness, although the height of the brick may vary from a nominal 2 in. to 6 in. *Firebrick* used for lining fireplaces and boilers are $9 \times 4\frac{1}{2} \times 2\frac{1}{2}$ in. Firebrick are made larger in each dimension to compensate for the absence of a mortar bed. Such brick are dipped into a thin fireclay mortar. Figure 7.1 shows the triangular dip marks on the light-colored brick of the dome of the kiln.

Common brick are produced with a smooth surface. Face brick, however, are often given a surface texture, such as rug, bark, stipple, or matt finish (Fig. 7.3).

The strength of a brick depends greatly on its porosity and therefore on the extent of vitrification obtained in the burning process. A low-quality common brick may have a porosity as high as 55 per cent, and a high-quality face brick a porosity as low as 10–15 per cent. The low-quality brick has the appearance of not having been fired very thoroughly, and appears relatively soft and weak. The more handling of such softer brick on the jobsite results in considerable breakage.

Highly porous brick also have a low resistance to penetration of rainwater and to freezing action; indeed, their resistance in these respects may be far less than that of mortar or even lightweight concrete. Spalling and cracking of brickwork by frost action is less severe in cold climates than in climates where winter temperatures fluctuate above and below freezing.

ASTM C62-66 defines three grades of common brick:

1. *Grade SW:* for use where a high degree of frost resistance is required, as in wet locations or below grade.

2. *Grade MW:* for use where exposed to temperatures below freezing but unlikely to be permeated with water, or where a limited degree of frost resistance is desired.

3. *Grade NW:* for use as interior or backup masonry, where frost resistance is not a requirement.

Minimum compressive strengths (psi; brick flatwise) are the following:

| | Average of Five Bricks | Individual Brick |
|---|---|---|
| SW | 3000 | 2500 |
| MW | 2500 | 2200 |
| NW | 1500 | 1250 |

## 7.3  bond patterns

The pattern of layup of brick in a wall is known as *bond* or *pattern bond*. The familiar overlap from one course to the next is known as *half-bond*. Stack bond is shown in Fig. 6.7.

Brick has three dimensions and three sizes of surface. Any of the three surfaces may be presented to the exterior, resulting in the *stretcher*, *header*, and *soldier courses* shown in Fig. 7.4. The *rowlock course* is another possibility, also shown in the figure. If stretcher courses are used for both exterior

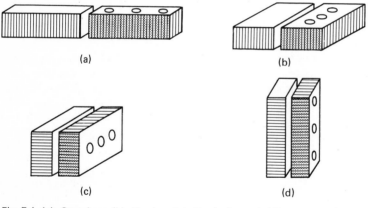

(a)

(b)

(c)

(d)

Fig. 7.4 (a) Stretcher; (b) Header; (c) Rowlock; and (d) Soldier courses.

and backup brick in an 8-in.-thick wall, a header course will bind together the two layers of the brick wall. In the case of a cavity wall (Fig. 7.5) there is a cavity or separation between the inner brick wall and the exterior veneer brick. The two components of the wall must still be bonded together for resistance to movement or collapse; this is done by metal ties, as shown in the figure. Brick veneer over a wood wall is supported by a nailed metal strap (Fig. 7.6). Bonding to a backup wall of tile or of concrete block is shown in Fig. 7.7.

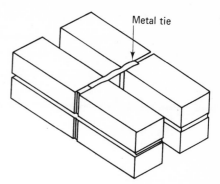

Fig. 7.5 Cavity wall reinforced with metal ties.

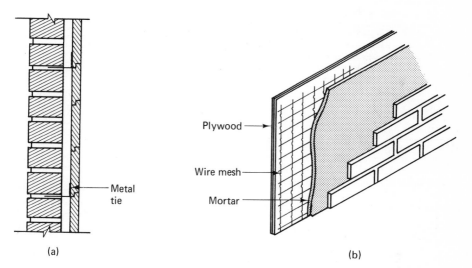

(a)

Plywood

Wire mesh

Mortar

(b)

Fig. 7.6 Brick veneer on wood sheathing: (a) using metal ties; (b) using mortar.

Fig. 7.7 Bonding face brick to backup masonry units.

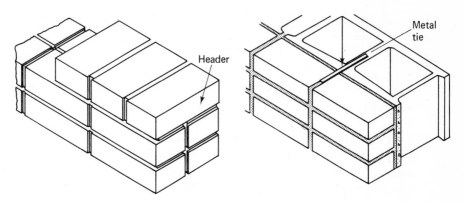

Header

Metal tie

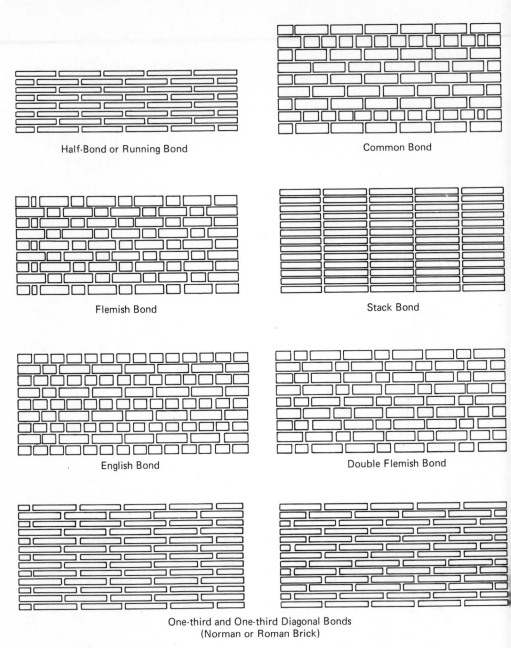

Half-Bond or Running Bond

Common Bond

Flemish Bond

Stack Bond

English Bond

Double Flemish Bond

One-third and One-third Diagonal Bonds
(Norman or Roman Brick)

Fig. 7.8 Bond patterns for brick.

    Five basic brick bond patterns are presently in common use, shown in Fig. 7.8. *Running bond*, which is the familiar half-bond, is a series of stretcher courses, for cavity wall and brick veneer construction. Metal ties are required for stabilizing a wall in running bond. *Common bond* is a running bond pattern with a header course every fifth or sixth course to provide structural strength.

The *Flemish bond* alternates a header and a stretcher brick along each course. The header is centered over the stretcher brick in the course below. If full-length headers are not required to tie into the backup course of brick, a half-length header or blind header of half a brick is used. *English bond* alternates a course of stretchers and a course of headers. The headers are centered on the stretchers and the joints of the course below. In *English cross bond*, sometimes called *Dutch bond*, header and stretcher courses alternate, but each stretcher course is displaced a half-bond from the next stretcher course.

Longer brick such as Norman offer the possibilities of a one-third bond, shown in Fig. 7.8.

## 7.4 structural clay tile

Brick are considered to be solid masonry units. Some types of brick may have perforations, but these must not exceed in area 25 per cent of the total cross-sectional area. Structural clay tile are hollow masonry units with cores usually rectangular in section, either horizontal or vertical, and are burned like brick. Because of their hollow shape, structural clay tile must be formed by extrusion in the stiff mud process.

Structural burned clay tile serve a number of functions in building construction:

1. Load-bearing wall tile for the construction of exposed or faced load-bearing walls. The facing material may be stucco, plaster, stone, or other veneering material.
2. Partition tile, a non-load-bearing application.
3. Backup tile for construction of composite walls of brick or other material backed up by tile, with the total load supported by both the facing material and the tile.
4. Furring tile, a non-load-bearing application.
5. Fireproofing tile, a non-load-bearing application, the tile providing fireproofing for steel members of the building frame.
6. Floor tile.

Hence there are two basic classes of tile, load-bearing and non-load-bearing. Non-load-bearing types must meet only one physical property specification, which is a limitation on water absorption rate (ASTM C56-62, Part 12). Load-bearing tile has stricter limitations on water absorption and must meet compressive strength requirements. The meaning of "load-bearing" here is that the masonry wall must carry other building loads besides its own weight. A non-load-bearing wall carries only its own weight.

## 7.5  load-bearing clay wall tile

These units are designed for use in bearing walls of light building construction, the height being governed by the local building code. Such tile may be exposed or faced with a veneering material such as stucco. ASTM Specification C34-62 defines two grades, LBX and LB. The higher-grade LBX is suitable for general use and for masonry exposed to weathering. Grade LB is suited for general use if not exposed to frost action, or for use in exposed

Fig. 7.9 Various burned tile shapes.

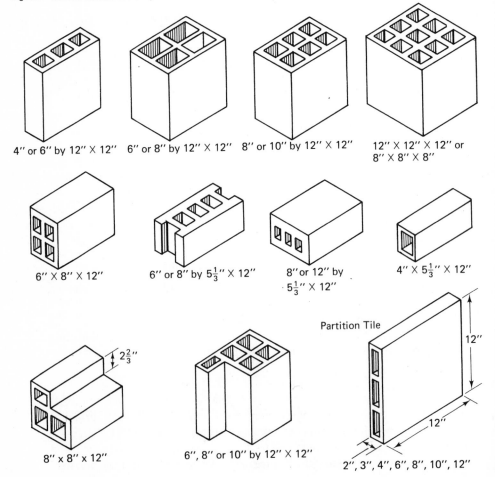

4″ or 6″ by 12″ X 12″    6″ or 8″ by 12″ X 12″    8″ or 10″ by 12″ X 12″    12″ X 12″ X 12″ or 8″ X 8″ X 8″

6″ X 8″ X 12″    6″ or 8″ by 5⅓″ X 12″    8″ or 12″ by 5⅓″ X 12″    4″ X 5⅓″ X 12″

2⅔″

8″ x 8″ x 12″    6″, 8″ or 10″ by 12″ X 12″

Partition Tile

12″

12″

2″, 3″, 4″, 6″, 8″, 10″, 12″

masonry when protected with at least 3 in. of stone, brick, or other masonry. Water absorption and strength specifications for LB grade are somewhat less stringent than those for LBX grade, as shown in the following table:

### MINIMUM COMPRESSIVE STRENGTH BASED ON GROSS AREA (psi)

| | End Construction Tile | | Side Construction Tile | |
|---|---|---|---|---|
| Grade | Average, 5 tests | Individual tile | Average, 5 Tests | Individual Tile |
| LBX | 1400 | 1000 | 700 | 500 |
| LB | 1000 | 700 | 700 | 500 |

Structural load-bearing tile are made in the following thicknesses: 4, 6, 8, 10, and 12 in. Some of the standard shapes are shown in Fig. 7.9. The nominal sizes include the thickness of the mortar joint for all dimensions.

Backup tile may be built into either load-bearing or non-load-bearing walls and are faced with either brick or facing tile. The face units are bonded to the backup tile so that both carry the loads in an integral wall construction. Bonding methods are shown in Fig. 7.10. Modular thicknesses are 4, 6, and 8 in. Nominal sizes includes the thickness of the mortar joint. Both header and stretcher tile, in both vertical and horizontal cell, are produced.

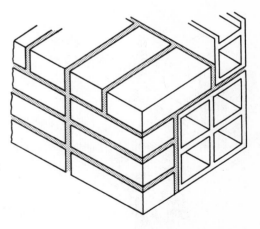

Fig. 7.10 Backup tile with brick veneer.

## 7.6 non-load-bearing clay tile

Typical partition tile are shown in Fig. 7.9.

Furring tile are made for the inside surfaces of exterior walls. Their purpose may be to provide a surface for plastering, to provide a moisture barrier for the wall, or to offer air spaces for heat insulation. Both solid and split types are illustrated in Fig. 7.11. Solid tile are available in 2-, 3-, and 4-in. thickness. The split tile is a half-section solid tile. Typical furring tile installations are illustrated in Fig. 7.12.

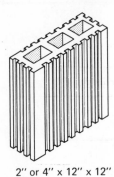

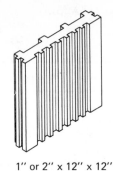

Fig. 7.11  Solid and split furring tile.

2" or 4" x 12" x 12"

1" or 2" x 12" x 12"

Although steelwork does not burn, it is not resistant to the effects of elevated temperatures. Open steelwork is frequently covered to protect it from fire. One of the many fireproofing materials is fireproofing tile. Partition and furring tile may be used for this purpose in any suitable shape or size,

Fig. 7.12  Installation of solid and split furring tile.

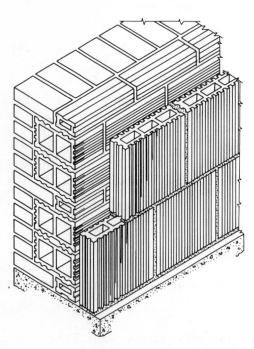

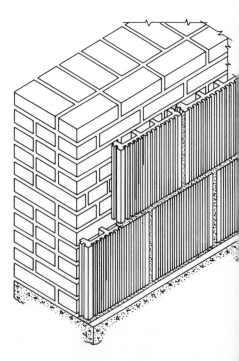

while soffit, angle, and clip types of tile are suited to the contours of rolled structural shapes, as shown in Fig. 7.13.

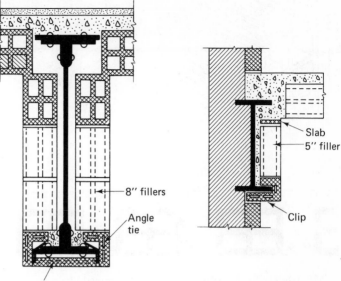

8" fillers

Angle tie

Slab
5" filler

Clip

2"-soffit

Fig. 7.13 Fireproofing of structural steel with clay tile, including angle, clip, and soffit tile.

Different hourly ratings of fire exposure are given to different types of fireproofing, as illustrated in Fig. 7.14.

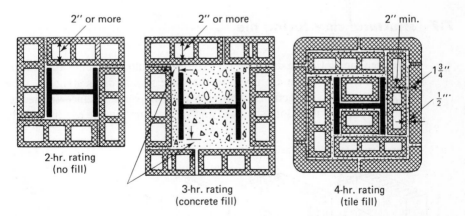

2" or more

2" or more

2" min.

$1\frac{3}{4}''$

$\frac{1}{2}'''$

2-hr. rating
(no fill)

3-hr. rating
(concrete fill)

4-hr. rating
(tile fill)

Fig. 7.14 Fireproofing ratings using tile.

Ribbed concrete roof and floor slabs may incorporate clay floor tile in their construction as in the example of Fig. 7.15. Structural clay floor tile are

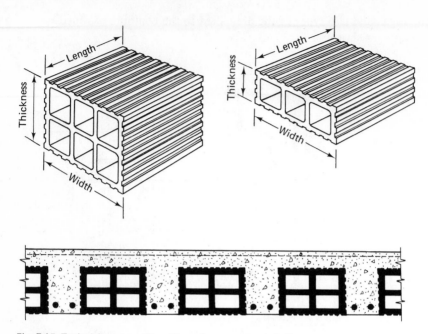

Fig. 7.15 Typical floor or ceiling tile and a concrete and tile ceiling.

produced in both load-bearing and non-load-bearing grades in thicknesses from 3 to 12 in. Standard length and width is 12 in., although other sizes are also used.

## 7.7 structural clay facing tile

Unglazed facing tile are designed for exterior and interior facing applications. Special-duty facing tile are made to the same standards as standard facing tile, but have heavier sections to provide better impact resistance and reduced moisture penetration. Both standard and special-duty tile are produced in two

Fig. 7.16 Some shapes of unglazed facing tile.

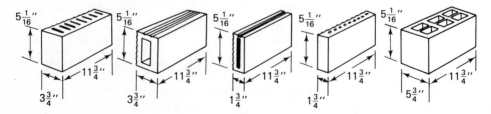

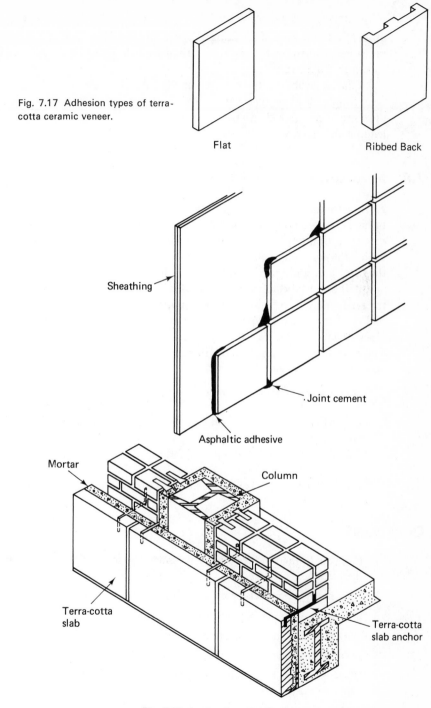

Fig. 7.17 Adhesion types of terra-cotta ceramic veneer.

Flat

Ribbed Back

Sheathing

Joint cement

Asphaltic adhesive

Mortar

Column

Terra-cotta slab

Terra-cotta slab anchor

Fig. 7.18 Application of adhesion and anchor types of terra-cotta veneer.

types, FTX and FTS. Type FTX is a smooth tile, designed for low moisture absorption and ease of cleaning. Type FTS may be smooth or textured. Its moisture absorption rate is higher than that of FTX tile. Some shapes are shown in Fig. 7.16.

Glazed structural facing tile is made with a ceramic or salt glaze on the face. Single-faced or type I units are glazed on one face; type II units have opposite faces glazed. Shapes and sizes are generally restricted to 16-in. lengths and narrow thicknesses.

## 7.8 terra-cotta

*Terra-cotta* means "fired earth," and this definition could apply to all types of burned clay products. This is a decorative burned clay unit used as a ceramic veneer, with a flat face and ribbed or flat back (Fig. 7.17). It is produced in both adhesion and anchor types, for mortar bond to the backup material in the first case, and mortar and wire ties in the case of the anchor type.

Adhesion types of terra-cotta are available in lengths up to 36 in., with thickness limited to $1\frac{5}{8}$ in. Methods of application are suggested in the details of Fig. 7.18. Anchor types may be made in larger slabs than the adhesion types of terra-cotta, and the thickness is somewhat greater, perhaps $2\frac{1}{2}$ in. The back of an anchored slab will be ribbed or scored for adhesion and anchor holes are provided.

### PERTINENT ASTM SPECIFICATIONS

C34-62 (Part 12) Structural Clay Load-Bearing Tile
C62-66 (Part 12) Building Brick
C67-66 (Part 12) Sampling and Testing Brick
C216-66 (Part 12) Facing Brick
C43-65T (Part 12) Terms Relating to Structural Clay Products

## QUESTIONS

1   Can you distinguish between dry press and pressed brick by the appearance of the brick?

2   Why must green brick be dried slowly?

3   Sketch header, stretcher, soldier, and rowlock courses.

4   Why are brick soaked before laying up in a wall?

5   Distinguish between brick and tile.

6   What is the basic difference between the requirements imposed on face and common brick?

7   Collect a variety of types of common brick and determine the modulus of rupture of each.

Mortars are cementing mixtures of materials for the joining of brick, stone, block, and other ceramic wall units. In selecting the mortar, seven factors are considered:

1. Mortar strength.

2. Appearance.

3. Water and rain penetration.

4. Cost.

5. The bond between mortar and brick.

6. Workability, which is the flow characteristic of a mortar which enables it to fill the joint completely.

7. Water retentivity, the ability of the mortar to retain water and prevent it from escaping into the masonry units.

# MORTAR
# JOINT

# 8

## 8.1 types of mortars

Four types of mortars serve most building construction requirements:

1. Cement mortars.
2. Cement-lime mortars.
3. Lime mortars.
4. Plaster mortars.

Plaster mortars, used in plastering operations, contain gypsum, which is not a basic ingredient in brick mortars.

Cements were discussed in Chapter 5. In addition to the standard portland cement, masonry cement is designed specially to provide a better mortar than one made up with standard portland cement or a lime-cement mortar. Masonry cements are composed of portland cement clinker, calcium limestone, gypsum, and an air-entraining agent. A plasticizing agent and a set retarder or accelerator may also be incorporated in the masonry cement. The limestone and air-entraining agent assist in providing plasticity and workability. The air-entraining agent also contributes to the ability of the mortar cement to retain its water, thus increasing adhesion to masonry units which tend to draw water from the mortar.

As explained in Section 4.7, lime hardens slowly by combining with carbon dioxide in the atmosphere.

Sand is required in mortar to reduce cost and especially to reduce shrinkage. As a rule of thumb, the volume of sand should be three times the volume of cement plus lime in the mortar. Water for mortar must be clean and must not contain excessive amounts of acid, alkali, or organic material.

The straight lime mortars are more plastic; the cement mortars are stronger. Portland cement provides poor plasticity but excellent compressive strength and faster setting.

The strength of a masonry wall depends upon the strength of both the masonry units and the mortar. A strong mortar would not be employed to join weak masonry units. The strongest mortars are made with cement, but these have the disadvantage of shrinkage cracking.

Penetration of water and rain occurs chiefly through cracks, including shrinkage cracks, in the masonry and mortar. Shrinkage in the mortar must therefore be reduced to the least amount possible, although there is not much difference in shrinkage between the various mortars in practice. Cracks occur initially at the contact area between masonry and mortar joint, indicating that good adhesion is an important characteristic of the mortar. Shrinkage can also occur in the masonry units, especially in the case of concrete block and sand-lime brick. If conditions permit, weak mortars should be used under these circumstances. A weak mortar will distribute the total amount of shrinkage over a large number of hairline cracks rather than concentrating the shrinkage strains into a single large crack. At weak points in a wall, such as window and door openings, cracks are more likely to occur. The reinforcing of the mortar joint with expanded metal provides some control over shrinkage in these areas.

Adhesion of the mortar to the masonry is improved by slight wetting of the masonry surface before laying, especially in dry weather.

The following four types of mortar serve most requirements in masonry construction.

1. *Type M Mortar:* This is a general-purpose mortar, giving high early strength and high ultimate strength, and is recommended for damp conditions such as masonry below grade or in contact with earth. It is made up of the following proportions by volume:

> 1 part portland cement
> 3 parts sand
> $\frac{1}{4}$ part hydrated lime or lime putty for workability

or, alternatively,

> 1 part portland cement
> 1 part masonry cement
> 6 parts sand

2. *Type S Mortar:* Type S is another general-purpose mortar especially recommended for resistance to lateral forces. Its ingredients are, by volume:

> 1 part portland cement
> $4\frac{1}{2}$ parts sand
> $\frac{1}{2}$ part hydrated lime or lime putty

or, alternatively,

> $\frac{1}{2}$ part portland cement
> 1 part masonry cement
> $4\frac{1}{2}$ parts sand

3. *Type N Mortar:* This mortar is suited to exposed masonry above grade or subject to severe and wet weather and freezing. It is not recommended for concrete block with a high shrinkage. It has the following proportions by volume:

> 1 part portland cement
> 1 part hydrated lime or lime putty
> 6 parts sand

or, alternatively,

> 1 part masonry cement
> 3 parts sand

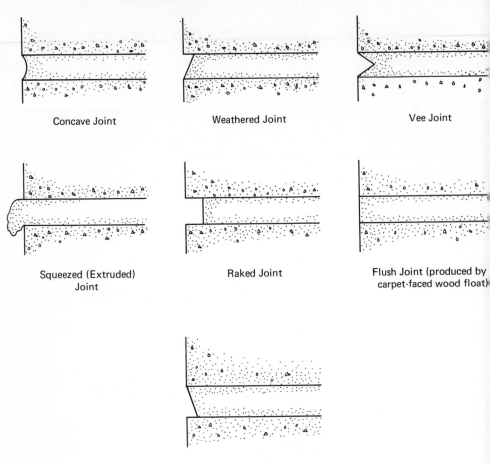

| Concave Joint | Weathered Joint | Vee Joint |
| Squeezed (Extruded) Joint | Raked Joint | Flush Joint (produced by carpet-faced wood float) |

Struck Joint (used when plaster
or other finish is to be applied)

Fig. 8.1 Mortar joint treatments. The concave, weathered, and
vee joint are highly resistant to moisture penetration. The other
types shown are not.

4. *Type O Mortar:* This type is used for load-bearing walls, where stresses
do not exceed 100 psi and the masonry is not subject to freezing. The
mixture is, by volume:

> 1 part portland cement
> 2 parts hydrated lime or lime putty
> 9 parts sand

or, alternatively,

> 1 part masonry cement
> 3 parts sand

Various treatments of the mortar joint are shown in Fig. 8.1. The tooled joints give best resistance to penetration of water, since they press the mortar against the brick.

Reference to a mortar bed standard thickness of $\frac{3}{8}$ in. in the dimensioning of brick and concrete block was made in previous chapters. It is usually assumed that the strength of the masonry joint is decreased by about 15 per cent for every $\frac{1}{8}$-in. increase in the joint thickness, although this rule of thumb probably has not been proved. Thicker joints may well be weaker because of the increased susceptibility to racking, or the rounding of joints illustrated in Fig. 8.2.

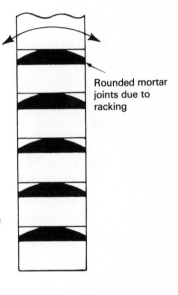

Rounded mortar joints due to racking

Fig. 8.2 Racking or rounding of mortar joints.

## 8.2 adhesion to the masonry unit

The tendency of masonry units to absorb water from the mortar is called *suction*. This absorption of masonry is measured in several ways: absorption in a 24-hr submersion test and absorption in a 5-hr boiling test are two specified by ASTM. Perhaps the most convenient is the initial rate of absorption (I.R.A.) defined by ASTM 67 as the amount of water initially absorbed by a dry masonry unit when it is immersed in water to a depth of $\frac{1}{8}$ in. for a time of 1 min. This test corresponds most closely to the brickmason's concept of suction. The I.R.A. is measured in grams per minute per 30 in.$^2$, and the values for masonry units can range from 1 to more than 50. Denser masonry units show lower suctions.

The mason overcomes severe suction by wetting the masonry unit before laying it. Concrete block should not be wetted, however, because of the increased shrinkage that will result from wetting.

A good mortar bond therefore depends on the I.R.A. or suction of the masonry unit and the water-retention characteristics of the mortar mix. Mortar with a low water retentivity used with masonry units of high suction could produce a condition of mortar dry-out, which would result in an inferior bond.

A field test for suction is to draw a circle the size of a 25-cent piece with a grease pencil on the face of the masonry unit. Place about 20 drops of water in the circle. If all the water is absorbed by the masonry unit in $1\frac{1}{2}$ min, the brick should be wetted before laying. Concrete block and sand-lime brick should not be wetted, however.

## PERTINENT ASTM SPECIFICATIONS

C270-64T (Part 12) Mortar for Unit Masonry
C190-63 (Part 12) Tensile Strength of Hydraulic Cement Mortars

## QUESTIONS

1   What is the difference in the chemical method of hardening between slaked lime and masonry cement?

2   What are (a) suction; (b) retentivity of a mortar?

3   Why does a mortar require sand?

4   Why is retentivity of a mortar an important characteristic?

The bonding materials for plaster mixes are portland cement, lime, and gypsum. Gypsum is a soft hydrous calcium sulfate, $CaSO_4 \cdot 2H_2O$. Gypsum products have excellent fire resistance due to slow release of the water of crystallization ($2H_2O$) in the gypsum.

Most plastering jobs require either two coats or three. Plaster applied over masonry, wood lath, or gypsum lath usually requires only two coats, while metal lath usually requires three. The several coats are termed *scratch coat* (first or base coat), *brown coat* (second coat), and *finish* or *putty coat*. In a two-coat system the first coat is a combined scratch and brown coat. Before the scratch coat sets, its surface is raked or scratched, usually in two directions, to produce an adequate bond for the following brown coat. The brown coat is about $\frac{1}{2}$ in. thick and is not as rich a gypsum mortar as the scratch coat.

There are a rather large number of types of plasters.

*Unfibered gypsum* is the neat product, sometimes called cement plaster. Neat plaster has no aggregate. Water and aggregate are added on the job for scratch and brown coats. There are three types of this grade:

# GYPSUM
# AND
# PLASTER

# 9

1. *Regular:* for use with sand as aggregate and hand application of the plaster.

2. *LW:* for use with lightweight aggregates such as perlite, and hand application.

3. *Machine application:* for sand or lightweight aggregate, and machine-applied.

*Fibered gypsum* is a neat gypsum with cattle hair or fiber added to it. When used with metal lath, the fibers reinforce the plaster and prevent too much plaster being pushed through the mesh openings. Fibered plasters are not suitable for machine application because of the problems of passing fibers through the nozzle of the machine. *Wood-fibered gypsum* uses wood fibers.

*Hard wall plaster* is a neat gypsum plaster containing hair or fiber and used for scratch or brown coats. Water and aggregate are added on the job, the aggregate being either sand or lightweight perlite or vermiculite.

*Bond coat gypsum* is specially prepared for use as a base coat on concrete surfaces that are smooth with insufficient suction and bond for ordinary plaster bases. Only water is added.

*Plaster of paris* is a calcined gypsum, $CaSO_4 \cdot \frac{1}{2}H_2O$. When mixed with water it sets in about 20 minutes. Plaster of paris is a material of many uses. Within the construction industry it is used for patching plaster walls and for ornamental plastering. Mixed with lime putty it gives a finish coat with low shrinkage and fast hardening.

*Keene's cement* is another calcined gypsum. Regular (slow-setting) and quick-setting types are available. Keene's cement makes a high-strength and hard plaster with slaked lime and sand, providing impervious, smooth surfaces resistant to moisture.

*Finish plaster* is used for the finish or putty coat. This grade of plaster is mixed with slaked lime putty and water, in the proportions 1 part plaster to 3 parts lime putty by volume.

*Fireproofing plasters* are sold under brand names for direct spray application to bare steel shapes. They are neither decorative nor abrasion-resistant. They may contain asbestos, inorganic binders, and lightweight aggregates.

*Acoustical plasters* are designed to absorb acoustic energy. There are several compositions of acoustical plaster, based on either gypsum, lime, or Keene's cement, and incorporating rockwool, asbestos, pumice, perlite, or vermiculite. Most brands also incorporate a foaming agent to produce a porous plaster. These plasters require only water, and are then applied in a $\frac{1}{2}$-in. coat. The usual coverage is 4 yd$^2$ per bag.

*Joint fillers* are a type of finish plaster to fill nail holes and joints in gypsum wallboard. These are mixed with water.

Fig. 9.1 Rib lath with two meshes between ribs, and expanded metal mesh.

## 9.2 plaster lath

The surface, other than masonry, to which plaster is applied is called *lath*. Any suitable material that will offer sufficient bond to the plaster may be used as lath, including expanded metal mesh, wire mesh lath, wood lath, or various types of wallboard or plasterboard. Expanded metal mesh particularly offers a superior bond for plaster. Foamed polystyrene and polyurethane insulation boards have no suction but provide an adequate mechanical bond for plaster.

*Metal lath* for plaster (Fig. 9.1) is an expanded flat or ribbed type of mesh, $\frac{3}{8}$-in. size. Rib lath has ribs that run lengthwise of the lath in $\frac{1}{8}$-, $\frac{3}{8}$-, or $\frac{3}{4}$-in. rib height for plastering use. Corner beads are expanded metal angles used at external angles in walls to protect the plaster wall from damage and for corner reinforcement. Mesh may also need to be stiffened with $\frac{3}{4}$-in. steel channel, as in fireproofing the beam of Fig. 9.2.

Metal lath must be used where wall curves are of short radius and for superior fire ratings in lathing steel beams and columns.

*Lathing board* is also called gypsum lath. This is a gypsum core $\frac{3}{8}$ or $\frac{1}{2}$ in. thick covered on both sides with a heavy paper. The standard lath size is $16 \times 48$ in. In addition to this solid board, a perforated gypsum lath is produced with $\frac{3}{4}$-in. holes 4 in. apart in both directions for better bond. A third type is a solid board with reflective aluminum foil on one side for heat insulation and vapor barrier. For veneer plastering and drywall systems large boards 4 ft $\times$ 8 ft or up to 16 ft long are used to eliminate as many joints as possible. There is also available a special "type X" or Fireguard

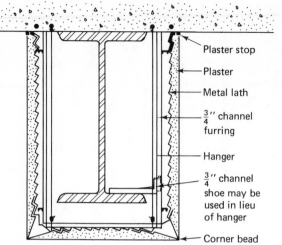

Plaster stop

Plaster

Metal lath

$\frac{3}{4}''$ channel furring

Hanger

$\frac{3}{4}''$ channel shoe may be used in lieu of hanger

Corner bead

Fig. 9.2 Lath and plaster for fire-proofing structural steel.

fire-retardant gypsum lathe, with a specially compounded core containing fire-retardant mineral fill.

Veneer plaster systems are designed to eliminate the brown coat of standard plastering sequences. These veneer plasters may be carried on standard gypsum board in two-coat systems or a veneer plaster gypsum lath in large sheets 4 ft wide, with a plain gypsum or a type X Fireguard core $\frac{1}{2}$ or $\frac{5}{8}$ in. thick. The veneer plaster is an extremely hard and strong plaster

Fig. 9.3 Taping joints in drywall construction.

with a total thickness of $\frac{3}{32}$ to $\frac{1}{4}$ in. These veneer plaster methods are particularly useful for embedded radiant-heat cables, since the thin coat of plaster can allow the heat to escape readily to the room without developing high enough temperatures to calcine the plaster. During installation, the joints and angles of the board are closed with a veneer plastering tape made of fiberglass mesh $2\frac{1}{2}$ in. wide. Veneer plaster is a kind of compromise between standard plaster systems and drywall construction.

Gypsum wallboard for drywall systems is somewhat similar to gypsum lath.

The smaller-size lathing boards are applied with their long edges parallel to the floor. Attachment is made with lath nails, staples, or patented clip systems. The nails are blued gypsum lath nails, $1\frac{1}{8}$ in. for $\frac{3}{8}$-in. boards and $1\frac{1}{4}$ in. for $\frac{1}{2}$-in. boards. Resin-coated nails and screw nails are also used.

Lime plaster mortars are not applied to gypsum lath or insulation board because of poor bonding, but may be applied to masonry surfaces.

## 9.3 gypsum wallboards, plank, and panels

*Gypsum wallboard* is a drywall or nonplastered type of sheathing applied to interior walls and to ceilings. Its construction is similar to gypsum lath, consisting of a gypsum core with facings of kraft paper. The paper on the exposed surface is an ivory color; the back surface is gray. The edges of the board are reinforced with paper. The sheets are 4 ft wide by 8 ft or more long, in $\frac{1}{4}$-, $\frac{3}{8}$-, $\frac{1}{2}$-, and $\frac{5}{8}$-in. thickness. The wallboard is applied in a single or double thickness of board, the doubled board usually being two thicknesses of board of $\frac{3}{8}$ in. each. A single thickness of board is preferably applied with the long edge horizontal. If double board is applied, the first board is applied vertically and the second horizontally, with a coat of gypsum cement joint filler applied to the back of the outer board. The outer board is nailed with double-headed nails until the joint filler is dry. The nails are then pulled and the nail holes filled with joint filler.

The joints of drywall board are reinforced with a paper tape pressed into the joint and coated with joint filler (Fig. 9.3).

Fig. 9.4 Gypsum roof plank.

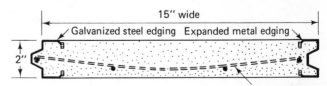

Galvanized steel edging  Expanded metal edging

15″ wide

2″

Galvanized wire mat

Roof plank made of reinforced gypsum is made in a number of styles. The tongue-and-groove type, 2 in. thick, shown in Fig. 9.4, is metal-edged; square-edged roof planks are made also.

Precast gypsum wall panels are made for exterior and interior use. They are usually 2 to 6 in. thick, 2 ft wide, and in lengths up to 10 ft. Edges are tongue-and-groove.

## 9.4 gypsum tile

Gypsum wallboard, roof plank, wall panels, and tile provide excellent fire ratings and low sound transmission. *Gypsum tile* is used in partition, furring, and fireproofing applications. *Fireproofing tile* is used to protect steel columns and beams. Shoe tile, angle tile, and soffit tile are made in sizes to fit the flanges of beams and applied as shown in Fig. 9.5. Standard 3-in. hollow gypsum tile is used to protect the webs.

*Partition tile* and *furring tile* (Fig. 9.6) include wood fiber for a strengthening binder and perlite aggregate for lightweight and thermal insulation. Both solid and hollow tile are manufactured, standard dimensions being

Fig. 9.5  Gypsum fireproofing tile for structural steel.

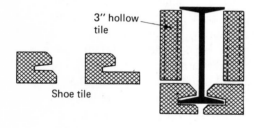

3" hollow tile

Shoe tile

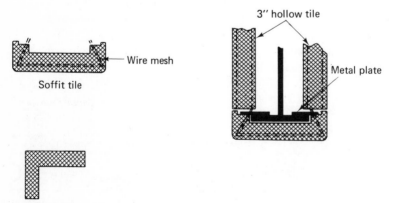

Wire mesh

Soffit tile

3" hollow tile

Metal plate

Angle tile

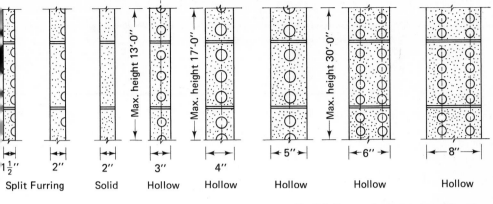

Fig. 9.6 Gypsum furring and partition tile.

12 × 30 in. in thicknesses of 2 to 6 in. Partition tile is either 2, 3, or 4 in. in thickness, 2-in. tile being adequate for partitions not exceeding 10 ft in height. These are a non-load-bearing tile.

## QUESTIONS

1 Name the coats in a three-coat plastering system.

2 What characteristic gives gypsum its excellent fire ratings?

3 What is fibered gypsum?

4 What is hard wall plaster?

5 What type of metal lath is a corner bead, and what purposes does it serve?

The architect Le Corbusier, a pioneer in the use of glass as an architectural material, once suggested that the story of architecture is simply the search for the window.

## 10.1  soda-lime glass

Many types of glass are produced, but only *soda-lime glass* is of significance for the construction industry. This type of glass, installed for window and plate glass, is a soda-lime silicate, and typically analyzes as follows (in per cent):

|  |  |
|---|---|
| silica | 74 |
| alumina | 1 |
| soda ($Na_2O$) | 15 |
| lime | 10 |

These proportions may be varied slightly. Fiberglass, for example, may have 10 per cent or more alumina with a reduction in silica.

# GLASS

# 10

Silica sand is the basis of glass. This is a remarkably pure white sand, virtually pure silica, usually consisting of fine grains of 50 to 100 mesh cemented together with kaolin. For glass manufacture a maximum iron oxide content of about 0.25 per cent is allowed, since the presence of iron oxide will give the glass a brown color similar to a beer bottle. Most of the silica sand mined for glass comes from the St. Peter formation, a sand deposit of vast extent in the central states. The formation extends from northern Michigan south to the Ozark area of Arkansas, but it is mined chiefly in Illinois, Minnesota, Missouri, and Arkansas. This silica sand was formed from Precambrian granites in Canada and exported duty-free to the United States by geologic movement in the Ice Age.

Silica and soda combine chemically to produce a glassy substance called sodium silicate (water glass), which is soluble in water and is used in large quantities as an adhesive in paper bag manufacture. If lime is added to this chemical reaction, the solubility of the product in water is reduced. When sufficient lime is added, a relatively insoluble glass product results, which may, however, still be attacked by water under certain circumstances of no interest to the construction industry.

Soda-lime glass, like most glasses, is not technically a solid but a supercooled liquid. A solid is a material that crystallizes from the liquid state at a definite melting temperature. When molten glass is cooled, it tends to crystallize, but its high viscosity prevents movements of the atoms to form crystals. The liquid state becomes "frozen in," making glass amorphous (noncrystalline). The amorphous condition is technically a liquid condition. Actually glass properties are unfavorably affected if crystals develop in the glass.

## 10.2 properties of soda-lime glass

Glasses are brittle materials, completely elastic to their ultimate tensile strength. They always fail in tension at relatively low stress levels of about 10,000 psi, because of the presence of innumerable submicroscopic surface cracks (see Fig. 10.1). However, freshly drawn glass fibers with extremely small diameters have given strength levels in tension of as much as $10^7$ psi. Any type of fiber glass is very much stronger than plate glass. The low strength of glass plate therefore requires care in the mounting methods and frame design of large glass curtain-wall construction, since stress levels sufficient to fracture glass are possible if the sheet of glass is restrained against movement.

The thermal conductivity of glass is high, the $U$ value for a single sheet being about 1 Btuh. To reduce heat transmission, *double-glazing* methods are used to take advantage of the insulation of the air space between panes. The air between the two glass plates must be dehydrated and the unit hermetically

Fig. 10.1 Familiar baseball effect on glass. Notice the two failure
circles and the radiating cracks. All fractures are tension failures.

sealed to prevent condensation on the inner surface of the glass. This type of
glass structure also reduces noise transmission. Double-glazing reduces the
$U$ value to about 0.54 Btuh.

Soda-lime glass transmits infrared and visible radiation but has a lower
transmission for ultraviolet radiation. Clean sheet glass $\frac{1}{8}$ in. thick transmits
about 90 per cent of visible radiation; thicker glass transmits less light. The
infrared radiation transmitted by glass is a source of building heat. If this heat
gain is excessive, the glass must be shielded from direct sunlight, and for this
purpose external shielding such as awnings or canopies is much superior to
internal shielding such as curtains. Special heat-absorbing and reflecting
glass is commonly used to reduce this solar heat gain. Heat-absorbing glass
is made by adding materials to the glass which absorb some of the solar
energy received by the glass. Such glass also has a lower transmission for
light and reduces glare in the room. This type of glass has a pale bluish-green
color.

## 10.3 special types of glass

*Wired glass* has a wire mesh inserted in the glass to prevent the glass from
shattering into dangerous shards under impact. Laminated safety glass has

the same safety characteristic but with a better appearance. Laminated safety glass is made from two thicknesses of sheet glass bonded with polyvinyl butyral, a transparent adhesive. This type of glass was first used, and is still used, for front windows of automobiles.

*Tempered plate glass* is a strong glass highly resistant to impact and thermal shock. It is made by a reheating and sudden cooling cycle that puts the outer surfaces of the glass under a residual compressive stress. As a result, the glass cannot readily break in tension as can an ordinary glass sheet. Tempered glass is used for swinging doors, sliding doors, and windows in sports arenas.

*Vitreous colored plate glass* is a strengthened and coated glass used for curtain walls and building fronts. The vitreous coat is a color coat fused to the glass surface. This is not a glazing material, but is applied against a masonry or other backup material.

## 10.4 glazing grades of glass

Sheet glass comes in three basic types: window glass, heavy sheet glass, and picture glass, with each type graded by quality as *AA*, *A*, *B*, or *Greenhouse grades*. The AA grade is specially selected glass for highest quality. A grade is superior. The B grade is acceptable for general glazing installations, and Greenhouse grade is used for greenhouse and similar horticultural and other nonwindow uses.

Window glass is made in single-strength and double-strength grades. Single-strength glass is 0.085–0.1 in. thick, 18 oz per ft$^2$; double-strength glass is 0.115–0.133 in. thick and 26 oz per ft$^2$. Single-strength glass in all qualities is produced in a maximum size of $40 \times 50$ in. The length plus the width of glass is referred to as "united inches," $40 \times 50$ being 90 united in. Double-strength glass in all qualities is produced in maximum size $60 \times 60$ or 120 united in.

Heavy sheet glass is supplied in $\frac{3}{16}$- and $\frac{7}{32}$-in. thickness for shelving, display cases, furniture tops, and similar applications, or for glazing applications where greater strength may be required. It is produced in AA, A, and B quality, and is available up to 120 in. long.

Picture glass is made in several thicknesses from 0.043 to 0.08 in. in AA, A, and B quality. This type of glass is used for picture framing, instrument dials, projector slides, and other miscellaenous uses.

Window glass is often a target for vandalism, especially when installed in city schools. Where such vandalism is costly, Plexiglas (acrylic) and polycarbonate plastic are substituted for glass. Acrylic plastic can be broken with some effort, while polycarbonate glazing is virtually indestructible by methods of brute force. These plastics are discussed further in Chapter 14.

## 10.5  glass block

*Glass block* is a masonry unit made of glass. Such blocks are made in two halves sealed together to enclose an air space for thermal and noise insulation. For installation, the edges of the block are coated with a mortar.

Glass block is produced in both functional and decorative types. They may be used to construct a glass masonry wall or may be combined with other masonry units in a wall (Fig. 10.2).

Fig. 10.2  Glass block construction for curved building contour.

Decorative blocks are produced in a range of surface textures, including etched, rippled, and fluted surfaces, and in clear or colored glass. Sizes are $6 \times 6$, $8 \times 8$, $12 \times 12$, and $4 \times 12$ in. Functional blocks differ from decorative blocks in that they are used to control and diffuse the light which passes through them to illuminate the building. Three styles are made:

1. A light-directing block bends incoming light upward toward the ceiling. This type is used above eye level (6 ft high) because if installed at lower heights, it could direct light into people's eyes.
2. A light-diffusing block diffuses light evenly throughout the room interior and thus may be installed at any level above the floor
3. A general-purpose block.

## 10.6 fiberglass and mineral wool

Fiberglass mats are made of fine filaments of glass, measuring about 20–75 $\times$ 10$^{-5}$ in. in diameter. A closely allied product is mineral wool, produced by a somewhat similar manufacturing process but employing the mineral diabase as the raw material. Both materials are available as loose insulating fill or as insulating blankets. In addition, fiberglass is woven as a reinforcing network to strengthen industrial paper products and fiberglass-reinforced plastics.

## QUESTIONS

1  What oxide in sand gives the sand a brown color?

2  Why is lime needed in glass?

3  Why is glass technically a liquid?

4  Glass is a hard material, and hard materials are strong materials. Why is the ultimate tensile strength of glass no higher than that of a hard plastic?

5  What methods are used to prevent glass from shattering?

6  Why is tempered glass stronger than ordinary window glass?

7  What is the meaning of "united inches".

8  To estimate the ultraviolet transmission of glass, place a piece of polyurethane foam outside a window and another inside a window. Ultraviolet degradation of the polyurethane is disclosed by the development of a brown color in the foam. Can you, by comparing the brown color developed in the two pieces of foam, estimate the ultraviolet transmission of the glass?

## 11.1  steel

The annual consumption of steel measured in pounds or tons is more than double the consumption of all other metals plus the plastics. After steel, the next most used of all manufactured materials is paper.

The remarkable variety of steels and the cheapness of steel explains the wide use of steel products. There is a steel for almost every purpose: soft and hard steels, ductile and brittle steels, magnetic and nonmagnetic steels, weldable steels and heat-treatable steels, wrought and cast steels, structural shapes and steel foil, and steels for the special service conditions of abrasion, hardness, heat, cold, corrosion, and impact. Notable, too, is the ease with which so many other metals can be alloyed with steel to provide an almost infinite variety of useful alloys.

A *steel* is a metal alloy based on iron, which includes also small amounts of carbon, manganese, and silicon. Steels that are alloyed with elements additional to these three are called *alloy steels*. Steels that do not include alloying elements other than small amounts of carbon, manganese, and silicon are called *carbon steels*.

# STEEL
# AND NONFERROUS
# METALS

# 11

Metals are rarely found in the natural state but must be extracted from their ores. Metal ores are actually corrosion products of metals. Iron rusts to red iron oxide, $Fe_2O_3$, which is the commonest ore of iron, called *hematite*. The black mill scale found on railroad rails, structural steel shapes, and hot-rolled steel plate is another oxide corrosion product, $Fe_3O_4$. The iron ore $Fe_3O_4$ is called *magnetite* because it is a magnetic ore. Hematite is not magnetic. These two oxides provide most of the world's iron and steel and are extracted in very large tonnages from large, open-pit mines.

It is not yet commercially feasible in the United States to convert iron ore directly into steel in a single operation. Such methods, termed "direct reduction" of iron ore, are in use only on a small scale. At present, the process from iron ore to steel requires two stages:

1. The ore is converted into hot metal (raw iron or pig iron) in a blast furnace.
2. The hot metal is refined and alloyed into steel in a steel refining furnace.

Fig. 11.1   Blast furnace in Pittsburgh.

The *blast furnace* (Fig. 11.1) is the most imposing, and the most complex, of all the big types of industrial furnaces, although the horizontal rotary cement kiln is longer. A blast furnace, of which there are about 200 in the United States, is a vertical shaft furnace about 200 ft high, with an inside diameter of 30 ft or more and a shape somewhat like an oversized Coca-Cola bottle. The construction of a blast furnace is a memorable contracting job, especially its foundations. Indeed, the blast furnace is now so expensive that probably only a few more will be built. It is not economically feasible to spend $60,000,000 or more to build a blast furnace that produces hot metal worth only a few cents a pound.

The production rate for a blast furnace is of the order of 3000 tons of hot metal per day, and the furnace must be in continuous operation without shutdowns. Raw material is charged into the top of the furnace by means of skip cars drawn up an inclined plane. Each finished ton of hot metal or pig iron requires the charging into the furnace of about 3350 lb of iron ore, 570 lb of limestone, and 1350 lb of coke, or a total of about 5300 lb of raw materials, not including $3\frac{1}{2}$ tons of air and about 40 tons of cooling water per ton of product. Nothing quite matches the impressive appetite of this furnace. The yearly material requirements reach rather awesome figures.

The limestone serves as a slag to remove some of the sulfur and phosphorus from the iron, two elements that contribute to brittleness in steel. The coke is made from coal in coke ovens, and burns to carbon monoxide. The carbon monoxide reduces the iron oxide to iron by taking oxygen from the ore at a temperature of about 3000°F to form carbon dioxide. The carbon found in all steels is derived from the coke originally charged into the blast furnace.

The hot metal is tapped out of the bottom of the blast furnace. This metal is actually an impure and brittle low-quality cast iron containing about $4\frac{1}{2}$ per cent carbon. Such a high carbon content makes the metal brittle and unusable as a steel. With so much carbon it has no ductility and cannot be rolled into finished shapes such as plate or structural shapes. This is the reason blast furnace iron must be processed further: among other requirements, the steel-refining furnace must reduce this carbon content.

The slag that floats on the molten metal in the blast furnace must also be tapped out of the furnace. As much of this slag as possible is used as construction aggregate for road surfacing, loose fill, and concrete products.

## 11.3 production of steel

Most steel is produced in the electric arc furnace or the oxygen converter. Considerable quantities of steel are still made in open-hearth furnaces, but open-hearth furnaces are slowly being phased out, to be replaced with oxygen converters. Although the open-hearth furnace will be in use for at least another twenty years, it will not be discussed here. It has been used since the beginning of modern steel-making practice about 100 years ago, and there is no lack of reference books on open-hearth practice for those interested in the subject. The Bessemer converter, which has disappeared from North America but still produces steel in Europe, will not be discussed here, either.

More than half of the total production of steel on this continent comes from the oxygen converter. The converter is a pear-shaped vessel open at the top. It has no external source of heat and therefore cannot melt metal. It is charged with 100 tons or more of hot metal plus some scrap steel. Then a retractable water-cooled lance is lowered to the bath of hot metal and high-pressure oxygen is directed into the bath. The stream of oxygen burns down the carbon in the iron to carbon monoxide at a rapid rate. This combustion reaction between carbon and oxygen is the only source of heat generated in the operation. As the temperature of the process increases, it frequently must be held back by additions of scrap steel. The open-hearth furnace and the electric furnace require a few hours to melt and refine a heat of steel; the oxygen converter can refine steel in about 20 minutes. It works so aggressively that it burns out its firebrick lining in a week of operation, and must be operated in pairs, with one converter producing and the other being re-bricked.

If molten hot metal must be transported from a blast furnace to an oxygen converter or open-hearth furnace, it is obvious that the converter or open-hearth furnace cannot be located more than a few miles from a blast furnace. The hot metal could be cooled in a mold and remelted in an open-hearth furnace, which has a burner, but this is an expensive method of operation. Hot metal should not be frozen only to be remelted again.

Steel is produced in Minneapolis and other cities where there are no blast furnaces. In the absence of a blast furnace, some other method of making steel must be employed that does not require hot iron. Steel can also be made from scrap. The electric arc furnace works on scrap steel. Specialty steels such as tool steels and stainless steels are made in the electric arc furnace, since such alloys must be produced in smaller tonnages than the oxygen converter can conveniently supply.

The *electric arc furnace* is rather like a very large electric arc welder. Electric arc welding and electric arc furnace operation are remarkably similar except in the quantities of metal melted. The arc welder operator melts weld metal with the heat of an arc. Slag is used to protect the deposited weld metal and to purify it, the slag being provided by a coating on the welding rod. The electric arc furnace uses three graphite electrodes about 12 in. in diameter to melt the charge of steel scrap. Two of the three white-hot electrodes may be seen projecting from the top of the furnace in Fig. 11.2. The third electrode is hidden behind the electrode on the left. Electric current travels down one electrode, arcs through the scrap steel, and returns to the transformer by another electrode. The scrap charge is dumped into the furnace by rotating the furnace roof and its electrodes off the furnace.

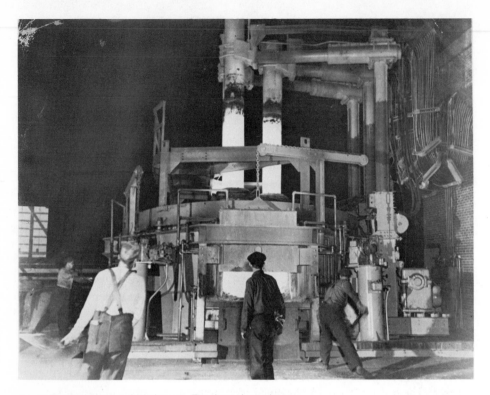

Fig. 11.2 Electric arc melting furnace. The three electrodes are partially withdrawn from the furnace while the furnace crew makes up slag. (Photo courtesy American Bridge Division, United States Steel Corporation.)

The arc furnace is made in sizes from $\frac{1}{2}$-ton capacity for small castings to 150 tons or more for steel production. In operation, the charge is dumped into the furnace, the roof is rotated back into position, the electrodes are lowered, and melting begins. When the scrap has melted, the refining operation begins. The carbon content is brought to the required low level, alloying elements are added, and a slag is made up to remove unwanted elements such as sulfur and phosphorus.

Besides alloy additions such as nickel, molybdenum, and chromium, two elements are required in small amounts in all steels. All steels require a minimum of about 0.3 per cent manganese. Manganese combines with any residual sulfur in the steel to form harmless manganese sulfide. Without manganese, sulfur in the steel would cause the steel to be brittle when hot and to crack during rolling-mill operations. Silicon, the other necessary additive, is used to deoxidize the steel. Dissolved oxygen in steel can make structural steel brittle at low temperatures. Sometimes aluminum is added to the steel, either as an alternative to silicon as a deoxidizer, or to make a fine-grained steel.

When the steel in the arc furnace has come to the correct carbon content as determined by the steel mill's metallurgical laboratory, the furnace is tilted to discharge the molten steel into a ladle carried from an overhead crane. From the ladle the steel is poured into ingot molds, where it freezes. These ingots of cast steel are rolled into the many steel shapes used by industry.

## 11.4 brittle fracture of steel

The tension test with its stress–strain data is a simple, reliable, and informative indication of the strength and ductility of a steel under most service conditions found in construction. Unfortunately there are certain circumstances that can cause a ductile steel to behave in brittle fashion. Two important cases of brittle failure of ductile steels can occur: brittle fracture and fatigue. The subject of fatigue failure can be only briefly noted here. A fatigue failure is a brittle failure in tension at a stress below the ultimate tensile stress, caused by a stress level repeated many thousands of times. Fatigue failures are the commonent type of failure in machine parts but are less commonly found in structural steelwork. On the other hand, there have been some dramatic failures of steel structures due to brittle fracture in the recent past. Such failures are now rare, because structural steels are produced that are less susceptible to this failure, and designers are more careful to design against brittle fracture. Some of the more spectacular brittle failures include bridges, storage tanks, pipelines, and ships, often fractured completely through the structure. Standard tension tests made on the fractured steel always reveal considerable ductility, indicating that this type of test is not suitable for disclosing susceptibility to brittle fracture. Figure 11.3 shows the difference between a ductile and a brittle failure in a structural steel.

Fig. 11.3 Ductile and brittle impact failures in a structural steel. The two samples on the left are brittle failures; the two on the right are ductile failures.

Under the conditions of a standard tension test the steel is subject to stress in one direction only. If a saw cut is now made in the tension specimen at right angles to the direction of stress, the steel will still maintain its ductility and will not be sensitive to the presence of this notch, even though at the root of the saw cut the stress conditions are two-dimensional and complex.

Under the impact conditions of the Charpy test, it is possible for the ductile steel to become remarkably brittle in tension. The *Charpy test* uses a notch in the specimen to produce a complex stress concentration at the root of the notch and is actually a test of notch sensitivity. It must be noted here that the other metals used in building construction, such as the aluminums, coppers, and stainless steels, are not affected by brittle fracture.

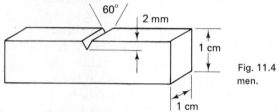

Fig. 11.4 Standard Charpy specimen.

For a Charpy test of a metal, samples of the shape and size shown in Fig. 11.4 are prepared. The figure shows a vee notch milled or broached across the specimen, but sometimes other types of notches are used, such as a keyhole shape. The specimen is supported horizontally in an impact testing machine (Fig. 11.5) and struck on the side opposite the notch by a swinging

Fig. 11.5 Impact tester with 60,000-lb Universal testing machine behind. The hammer is elevated to the 264-ft-lb position.

hammer that begins its swing with 264 ft-lb of energy. When the hammer strikes the specimen at the bottom of its swing, it has 264 ft-lb of kinetic energy. This energy is partly dissipated by fracturing the specimen, the remaining energy carrying the hammer up on the backswing. The Charpy machine indicates on a scale the amount of energy in foot-pounds required to fracture the specimen (actually it measures the backswing energy and subtracts this from the original 264 ft-lb to obtain the fracture energy).

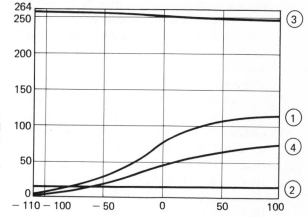

Fig. 11.6 Charpy impact test results: (1) notch-tough rimmed steel; (2) 2024 aluminum alloy in hard temper; (3) pure nickel; (4) 0.35-percent carbon steel plate $2\frac{1}{2}$ in. thick.

Suppose that a number of Charpy specimens of a particular steel are tested over a range of temperatures from $+150°$ to $-110°$F (temperatures down to $-110°$F can be obtained with xylene or acetone and pieces of Dry Ice). If the results are plotted on a graph of energy foot-pounds against temperature, a typical S curve results, similar to curve 1 of Fig. 11.6, which is the result for a notch-tough rimmed steel. At the high end of the temperature range the steel offers great resistance to fracture, giving a torn appearance at the break in what is termed a *ductile failure*. But as the temperature drops, a temperature range is reached in which the notch toughness of the steel falls drastically, the appearance of the fracture changes to a bright crystalline break, and the steel has become brittle.

The minimum Charpy impact strength allowed for many applications is 15 ft-lb per specimen. No steel should be used at a temperature giving a lower Charpy value than 15 ft-lb, and this temperature at which the energy absorption is 15 ft-lb. is often called the transition temperature between ductile and brittle failure under impact. Any of the following values might be termed the transition temperature for a brittle steel:

1. The temperature corresponding to 15 ft-lb.
2. The average temperature between brittle and ductile failure.
3. The inflection point of the S curve.
4. The lowest temperature at which the specimens show 100 per cent ductile failure at the break.

5. The temperature at which 50 per cent of the area of the break is ductile and 50 per cent brittle.
6. The highest temperature for which no part of the area of the break is ductile.

Brittle failure has been especially evident in welded structures. The thermal strains due to welding heat build a great deal of residual stress into structures, and these stresses can contribute to brittle failure. Such residual stresses are commonly as high as the yield stress of the steel.

Brittle failure occurs only below the transition temperature of the steel, and is prevented by using a steel with a lower transition temperature than any service temperature likely to be met. Good fabrication practice also is required: notches, arc burns, and other such defects serve as stress raisers where brittle fracture can commence.

Several methods are available for lowering the transition temperature of a steel. Brittle fracture is caused by certain small atoms dissolved in steel, carbon and oxygen chiefly. Carbon has the most pronounced effect; therefore structural steels have carbon levels below 0.25 per cent. Welding rods for structural steels do not contain more than 0.1 per cent carbon. Oxygen is removed by "killing" the steel with silicon or aluminum; this killing operation is discussed in the following section. Fine-grained steels will have lower transition temperatures than coarse-grained steels.

From Fig. 11.6 it is apparent that nickel has remarkable Charpy impact values. Nickel alloys well with steels, and to obtain low transition temperatures a little nickel is included in the steel analysis. A 3.5 per cent nickel low-carbon steel may be safely used at $-150°F$; with 9 per cent nickel the transition temperature is below $-320°F$. Manganese is also beneficial but does not have the powerful effect that nickel has. The nickel-bearing stainless steels contain about 9 per cent nickel and are completely notch-tough.

Thicker plates are more susceptible to this notch effect than thinner plates. Welding procedures for heavy structural plate must be set up with more care if brittle fracture is to be avoided. The steel mill must add a little extra carbon to thicker steel plate if it is to meet the strength of thinner plate, and this increment in carbon increases the risk of brittle fracture.

## 11.5 *rimmed and killed steels*

Plain carbon steels are made in three grades: rimmed, semikilled, and killed. A *killed steel* is deoxidized by addition of small amounts of silicon or aluminum to the ladle of steel. These two elements are strong oxide formers and remove the dissolved oxygen in the steel. Since there is no evolution of dissolved gas when the molten steel cools in the mold, such steels lie quiet in the mold ("killed"). A *semikilled steel* is partially deoxidized. A *rimmed steel* does not receive this deoxidation treatment and is named for the very characteristic rim of parallel crystals that grow inward from the mold as the steel freezes. Killed steels are more resistant to brittle fracture than rimmed steels and are commonly rolled into structural shapes and plate. Arc-welding rods are made of low-carbon rimmed steel.

## 11.6   rolling mill practice

After the molds are stripped from the ingots, the ingots of steel are reheated in gas-fired soaking pits to the required hot-rolling temperature. *Hot rolling* breaks down the coarse grains of the slowly cooled ingot to produce a fine-grained product. Hot-rolled products include structural shapes, bar, plate, concrete reinforcing bars, and sheet. A larger amount of sheet is rolled than of any other steel product. Sheet may be either hot-rolled or cold-rolled. Hot-rolled sheet is recognized by its coat of black mill scale. *Cold-rolled* sheet is given a final pass or passes through rolls at a lower temperature than is used for hot rolling. It has a bright finish but the cold-rolling process work-hardens the steel somewhat. As a result, cold-rolled sheet cannot be given the severe forming operations that a hot-rolled sheet can receive. The thinner gages of steel sheet are only produced in cold-rolled condition.

## 11.7   carbon in steel

The most important alloying element in steel is carbon. This element is very potent in its effect on the properties of steel, so few steels need more than 1 per cent carbon. Those steels in which carbon is the only significant alloying element (and ignoring the small amounts of silicon and manganese in all steels) are called *carbon steels*. Alloy steels contain small or large amounts of other alloying elements, principally nickel, chromium, and molybdenum.

Steels are designated according to their carbon contents in accord with the following classification:

| Type of Steel | % Carbon | Area of Use |
|---|---|---|
| low-carbon steel | 0.03–0.30 | structurals, plate, sheet |
| medium-carbon steel | 0.35–0.55 | machine parts |
| high-carbon steel | 0.60–1.5 | tools and tooling |
| cast irons | 2.5–3.5 | castings |

*Low-carbon steels*, also called *mild steels*, are produced in greater quantities than all other steels combined. Mild steel is soft, remarkably ductile, and therefore easily shaped. Another significant characteristic is its ready weldability. These characteristics make mild steels ideal for construction purposes.

The amount of carbon in mild steels is usually 0.15–0.20 per cent. This is too low a carbon content for heat treating the steel. A minimum carbon content of about 0.35 per cent is necessary if a steel is to be heat-treated or hardened. Hardenable steels are difficult to weld successfully, since the welding operation also hardens them. The use of a hardenable steel in a welded building frame would be dangerous, since any hard-welded areas could be brittle.

The *medium-carbon steels* find uses in the many kinds of machine parts: gears, shafts, pins, etc. Such machine parts often require to be hardened for wear resistance, hence the need for sufficient carbon to make these steels heat-treatable. The chief use for medium-carbon steels in construction is in reinforcing bars for concrete.

*High-carbon steels* are tool steels and are fabricated into chisels, drills, punches, dies, saw blades, etc. Tools require extreme hardness if they are to cut and form other metals. Tool steels are therefore formulated for maximum useful carbon content.

The cast irons are also alloys of iron and carbon, but since they are not rolled into shapes, they do not require manganese. The cast steel ingots that are rolled into structural shapes and plate are an entirely different metal from the cast irons, which resemble the hot metal or pig iron produced by the blast furnace.

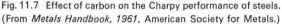

Fig. 11.7  Effect of carbon on the Charpy performance of steels. (From *Metals Handbook, 1961,* American Society for Metals.)

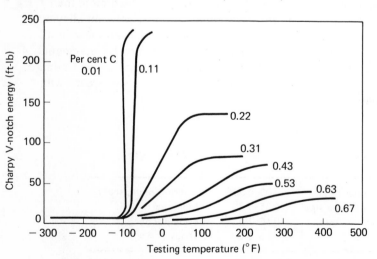

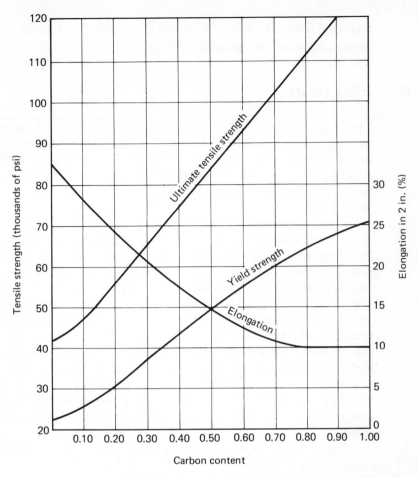

Fig. 11.8    Properties of hot-rolled steels as influenced by carbon content. The hot-rolled condition is the softest condition for these steels.

A small increase in the carbon content of a steel, even as little as $\frac{1}{10}$ of 1 per cent, has a strong effect on all the properties of the steel except the modulus of elasticity. If the carbon content is increased, these are some of the effects:

1. The melting point of the steel is lowered.
2. The steel is more easily heat-treated.
3. The steel becomes more difficult to weld without cracking.
4. The steel becomes harder.
5. The steel has a higher tensile strength.
6. The steel loses ductility.

Figure 11.7 shows the Charpy performance of steels at different carbon contents. This figure emphasizes again the need for low carbon in construction applications. In Fig. 11.8 the effect of carbon content on tensile and yield strength and on elongation is shown.

## 11.8 alloy steels

*Low-alloy steels* are those steels containing as much as 2–3 per cent total alloying metals, such as nickel, columbium, or copper. Such steels are used for applications that require either increased toughness or higher yield strength. *High-alloy steels* contain 5 per cent or more of alloying elements. Such high alloys find their applications under service conditions of wear, heat, or corrosion.

Until 25 years ago, the method of building great strength into equipment was the time-honored one of "beefing up," or making the equipment from heavier sections of steel. This method is often no longer practical. The method now preferred is to use a higher-strength low-alloy steel. Such a steel may offer, say, 30 per cent higher yield strength at an increased cost of 10 per cent over a steel of lower yield strength. Allowable loads will be increased by 30 per cent for a 10 per cent increase in material cost and no increase in labor cost. The high-strength low-alloy steel is therefore a bargain. Suppose that the use of the high-strength low-alloy steel permits a reduction of 2 in. in the depth of floor beams. In a tall building of 50 stories, the total building height is reduced by over 8 ft, which is a saving of many thousands of dollars. This represents 8 ft less of curtain wall, of fireproofing, of ducting, of insulation, of piping, and of installation labor, not to mention lower foundation costs.

The following major alloy steel groups are of greater or lesser significance for construction:

1. High-strength low-alloy structural steels (HSLA steels).
2. American Iron & Steel Institute (AISI) machinery steels, which are used in the machinery of construction equipment.
3. Stainless steels.

The AISI machinery steels are designated by a four-digit code that indicates the alloy and carbon content; e.g., 8620, 1040, etc. These steels are discussed in most steel handbooks and steel warehouse catalogues.

## 11.9 high-strength low-alloy

## steels

The standard structural steel used for decades was simply called "structural steel." It was made to ASTM Specification A7. This was a low-carbon steel, but the specification did not require the steel to meet any specified chemical analysis. Charpy impact values were always uncertain, and on occasion disastrous. The specification did, of course, require a certain tensile strength and ductility. Yield strength was 33,000 psi, ultimate tensile strength 60,000 psi, and elongation about 20 per cent in a 2-in. gage length. The design stress used with this steel was 20,000 psi.

The newer *high-strength low-alloy (HSLA) steels* have replaced A7. These are all "lean" steels, that is, very low in carbon and alloy content, with ultimate strengths that may be as high as 120,000 psi for some grades. By keeping alloy additions lean, the cost increment over A7 steel is held to a minimum and the steels do not create problems for the welder. To ensure toughness and weldability, the HSLA grades for construction purposes have carbon contents of 0.20 per cent or less. The alloy formulation is arranged to provide some resistance to atmospheric corrosion, and some grades may be left unpainted, the steel being protected by its own oxide. Of the wide range of brand names, perhaps the most well known are Cor-Ten and T-1. These are typical of the group. They contain small amounts of manganese, silicon, nickel, copper, and small amounts of either chromium, columbium, vanadium, or titanium for increased strength. Columbium is a popular alloy, since the addition of only 0.02 per cent columbium gives an increase in tensile strength of 10,000–15,000 psi.

Fig. 11-9 Construction in high-strength low-alloy steel. The building is an EXPO '67 pavilion.

Figure 11.9 shows a structure built of HSLA weathering steel. This alloy produces a tight coat of iron oxide that protects the steel from atmospheric corrosion. Hence it requires no painting, except for protection against severe environments. As the protective oxide develops, the color of this steel changes from orange to brown and finally to a permanent blue-gray color. The development of the surface oxide may be incomplete on interior steel if the water-vapor content of the building interior is low.

A minimum yield stress of 50,000 psi is specified for weathering steels. A typical analysis is 0.15 per cent maximum carbon, 1.35 per cent manganese, 0.20–0.50 per cent copper, 0.30–0.50 per cent chromium, and 0.25–0.50 per cent nickel. The combination of nickel with low carbon gives this steel impact toughness at low temperatures; the small addition of chromium provides increased tensile strength, and the copper provides the required corrosion resistance.

## 11.10   stainless steels

### for construction

The use of stainless steels for architectural panels, curtain wall, and trim is well established. These steels, although more expensive, have high strengths, good ductility, and complete resistance to all types of atmospheric corrosion. A stainless steel will maintain its original luster indefinitely, although it may require occasional washing to remove superficial grime.

The stainless steels may be summarized as very low carbon steels containing a minimum of 12 per cent chromium. The chromium provides the remarkable corrosion resistance of these alloys. There are two major groups of stainless steels, the 300 and the 400 series; all have three-digit numbers to designate the alloy, such as 304 and 430. The 400 series steels contain 12 per cent or more chromium and are magnetic. The 300 series steels contain usually 18–20 per cent chromium and about 10 per cent nickel and are non-magentic. The 400 stainless steels will not be discussed here, although they have occasional applications as construction materials; for example, stainless 430 is used as a chimney liner. However, most architectural stainless steels are from the 300 series.

All the 300 alloys are modifications of the basic alloy, which is 304. This is the standard low-carbon stainless steel, 0.08 per cent maximum carbon, with about 20 per cent chromium and 8–9 per cent nickel. The 304 alloy is a general-purpose stainless steel, easy to form and to weld. The other stainless steels of the 300 series have characteristics similar to 304. Alloy 302 has 0.12 per cent carbon, the increased carbon making for metallurgical difficulties in welding.

Fig. 11-10   Accessories in stainless steel, including the fire extinguisher.

Stainless fastenings must usually be used with stainless trim, since the use of other metals with stainless may lead to galvanic corrosion of the other metal in the presence of moisture. The 300 series steels have a thermal expansion one-third greater than standard steels, and there have been instances when this large expansion has caused difficulties. The high expansion may cause warping after welding, especially for thin gages of sheet.

A variety of finishes are available for stainless sheet, these finishes being designated by the numbers 1 to 8. Two types of finish are most commonly used:

1. *No. 2B:* cold-rolled, annealed, pickled in acid, then given a skin pass on polished rolls to produce a smooth, satin, mildly reflective finish. This finish is not too reflective for exterior panels.
2. *No. 4:* belt-ground with 150- to 180-grit abrasive to produce a polished finish.

## 11.11   structural steel shapes

Rolled steel shapes compete with reinforced concrete in carrying the dead and live loads of the building. The standard profiles for these shapes are I beams, wide-flange beams, channels, angles, tees, and zees (Fig. 11.11).

I *beams* and *wide-flange beams* are designated by their nominal depth, measured across both flanges, and weight per foot, thus:

S12 × 35: an I beam 12 in. deep, 35 lb per lineal foot

W24 × 100: a wide-flange beam 24 in. deep, 100 lb per lineal foot

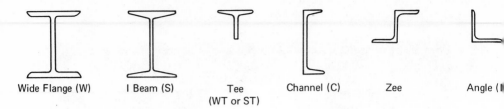

| Wide Flange (W) | I Beam (S) | Tee (WT or ST) | Channel (C) | Zee | Angle (L) |

Fig. 11-11    Rolled structural steel shapes.

The wide-flange or W beams, as the name indicates, have wider flanges than I beams (S beams). The wide-flange sections that are almost square in cross section, that is, as wide as they are deep, are an excellent shape for building columns. The flanges of W members have only a slight taper, while the flange of an I beam or S shape tapers with a slope of about 17 per cent.

*Channels* are also designated by depth and weight per foot, thus: C12 × 20.7. *Angles* are designated by the length of both legs and the thickness of metal in the legs, thus:

$$L2 \times 2 \times \tfrac{1}{8}$$
$$L6 \times 4 \times \tfrac{3}{8}$$

Channels and angles are chiefly used for the miscellaneous metal in buildings, such as angles for lintels and channels for the stringers of steel stairways.

Further details on the several types of rolled sections are readily available in handbooks and catalogues.

## 11.12    steel sheet and plate

*Steel sheet* is available either hot-rolled (mill scale) or cold-rolled (bright oiled finish), in various widths and lengths. Warehouse widths are 36 and 48 in., lengths 96, 120, and 144 in. The light gages, thinner than 20 gage (0.036 in.), are used for sheet metal work; the heavier gages, 16 gage and thicker, are found in miscellaneous metal applications. Since miscellaneous metal work in buildings must usually be welded, gages lighter than 16 gage are not generally used. Light gages will be warped or burned in manual arc welding operations. Even 16 gage is sufficiently light that it must be arc-welded with care; 14 gage is easy to weld.

The following table gives the thickness and weights of standard steel sheet. Only the even-numbered gages are shown, since the odd-numbered gages are not usually needed in building construction. To aid in recalling approximate thicknesses, note that 11 gage is often called $\tfrac{1}{8}$ in.; 16 gage is almost $\tfrac{1}{16}$ in., and 14 gage is almost $\tfrac{1}{14}$ in. The thicknesses of stainless steel and nonferrous sheets are slightly different in the same gage numbers: stainless and copper 16 gage measures 0.0625, for example.

Painted, porcelain-enameled, and vinyl-coated sheet are also used in construction.

*Steel plate* measures $\frac{3}{16}$ in. or greater in thickness and is hot-rolled. For estimating purposes, 1 ft$^2$ of $\frac{1}{4}$-in. steel plate weighs 10.2 lb., or nearly 10 lb:

### THICKNESS AND WEIGHT OF STEEL SHEET

| Gage | Thickness | Weight/Ft$^2$ |
|------|-----------|---------------|
| 10 | 0.1345 | $7\frac{1}{2}$ lb. |
| 11 | 0.1196 | $6\frac{2}{3}$ |
| 12 | 0.1046 | $5\frac{5}{6}$ |
| 14 | 0.0747 | $4\frac{1}{6}$ |
| 16 | 0.0598 | $3\frac{1}{3}$ |
| 18 | 0.0478 | $2\frac{2}{3}$ |
| 20 | 0.0359 | 2 |
| 22 | 0.0299 | $1\frac{2}{3}$ |

Interlocking sheet piling shapes are shown in Fig. 11.12.

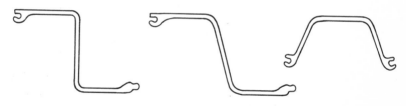

Fig. 11-12   Sheet steel piling.

### 11.13   pipe and tube

*Pipe* is used for light columns and other structural purposes as well as its primary use as a liquid conduit. *Tube*—round, square, and rectangular—is often employed for framing and light structural work.

The difference between pipe and tube is a complex thing to explain. It is best to ignore the complexities of the subject and to say that pipe is designed to be threaded for pipe fittings, whereas tubing is not threaded. Pipe is designated by its nominal inside diameter; round tubing is designated by outside diameter and wall thickness. For example, a 4-in. Schedule 40 pipe (Schedule 40 is pipe of standard wall thickness) measures 4.500 in. O.D., 4.026 in. I.D., while a 4-in. tube measures exactly 4.000 in. outside diameter. Perhaps it is best to think of tubing as hollow bar, whereas pipe is a conduit for liquids.

Both are available in various wall thicknesses, which are set out in handbooks.

The type of tubing used in construction is the type known as welded mechanical tubing. This type is not seamless but has a continuous seam weld along its length. Seamless tube is more expensive and cannot be justified in building construction.

## 11.14  corrugated steel pipe

Pipe over 6 in. in diameter is too cumbersome to thread. Large-diameter pipe is made of a wider variety of materials than small-diameter pipe, and these materials may be gray cast iron, steel, prestressed concrete, asbestos–cement, or combinations of materials such as cement-lined steel pipe.

Most pipe requires a rigid wall to resist internal fluid pressure. A different type of application is a pipe that must resist external soil or backfill pressures. Corrugated steel pipe serves this purpose (Fig. 11.13) and is most familiar in sewage and drainage applications such as culverts, although it has a wide range of other applications. Sewage and drainge service involves no internal fluid pressures, since the pipe rarely runs full.

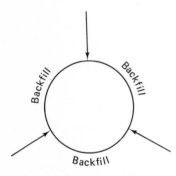

Fig. 11-13   Flexibility of corrugated thin-wall steel pipe is its strength. Backfill pressure supports the whole circumference of the pipe.

The wall of *corrugated steel pipe* (Fig. 11.14) is a corrugated light-gage steel sheet with a high degree of flexibility. If this flexible conduit is backfilled, earth pressure on the top of the pipe will cause the pipe to flatten, but the backfill will then react to provide side support to the pipe to resist such flattening. Earth loads therefore become uniformly distributed about the full circumference of the pipe. Since this type of pipe is usually buried, it is galvanized with 2 oz of zinc per square foot of sheet, that is, 1 oz on each side of the sheet.

The smallest diameter in which corrugated steel pipe is made is 8 in., available in 0.064- and 0.079-in. thicknesses; the largest diameter is 96 in. in a wall thickness of 0.168 in. Such drainge pipe is available also as pipe arch which is also shown in Fig. 11.14 with a bituminous-paved invert, and also

Fig. 11-14   Corrugated steel pipe. The pipe at the right is a pipe arch; both it and the middle pipe have a bituminous-paved invert.

asbestos-bonded. Asbestos-bonded pipe is made by pressing asbestos fiber into the molten zinc as the steel sheets come out of the galvanizing operation. The asbestos fibers are then saturated with bitumen. The sheets are corrugated and formed after these coating operations are completed.

## 11.15   reinforcing bar

*Reinforcing steel bar* for reinforced concrete is hot-rolled from either new steel or from scrapped steel axles of railroad cars. Rebar is supplied to ASTM mechanical requirements for tensile and yield strength, cold bend test, and elongation, in the three grades of structural, intermediate, and hard:

### MECHANICAL REQUIREMENTS FOR STEEL REINFORCING RODS

| Grade | Ultimate Tensile Strength (psi) | Yield Point (psi) |
|-------|--------------------------------|-------------------|
| structural | 55,000–75,000 | 35,000 min. |
| intermediate | 70,000–90,000 | 40,000 |
| hard | 80,000 min. | 50,000 |

Reinforcing bar is available as rounds, squares, and other shapes. Round bars are made both plain and deformed (Fig. 11.15), although most bar larger than No. 2 is deformed. The projections rolled into deformed bar provide better bond with the concrete. The allowable tensile load applied to

Fig. 11-15  Types of deformed reinforcing bar.

such a bar is governed by its cross section; the allowable bond strength to the concrete is governed by its perimeter or surface area. Bar sizes are designated by numbers, the number indicating the number of eighth-inches in the diameter. Thus a No. 3 bar is $\frac{3}{8}$ in. and a No. 7 is $\frac{7}{8}$ in. in diameter. Most bar used is in the sizes from 2 to 11, although larger bar is made:

### REINFORCING BAR

| Number | Diameter | Pounds/Foot | Cross Section (in.$^2$) | Perimeter (in.) |
|--------|----------|-------------|-------------------------|-----------------|
| 2 | 0.250 | 0.167 | 0.05 | 0.786 |
| 3 | 0.375 | 0.376 | 0.11 | 1.178 |
| 4 | 0.500 | 0.668 | 0.20 | 1.571 |
| 5 | 0.625 | 1.043 | 0.31 | 1.963 |
| 6 | 0.750 | 1.502 | 0.44 | 2.356 |
| 7 | 0.875 | 2.044 | 0.60 | 2.749 |
| 8 | 1.000 | 2.670 | 0.79 | 3.142 |
| 9 | 1.128 | 3.400 | 1.00 | 3.544 |
| 10 | 1.270 | 4.303 | 1.27 | 3.990 |
| 11 | 1.410 | 5.313 | 1.56 | 4.430 |

A variety of accessories are used to hold reinforcing bars in proper position, both vertically and horizontally in the forms. Some of these are shown in Fig. 11.16.

A spiral-wound reinforcing bar made from intermediate or hard grades of steel is available for the reinforcement of concrete columns. This spiral bar is available in rod diameters Nos. 2, 3, 4, and 5, in spiral diameters through the range 10–40 in. For installation, the coils of the spiral are supported in required spacing in the column by spacer pieces.

For reinforcing flat areas of concrete such as slabs, reinforcing mesh is used. The mesh is available in various gages of wire woven longitudinally and transversely at different intervals. One-way mesh has rectangular openings

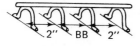

Slab Bolster

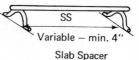

Slab Spacer

Bar Chair

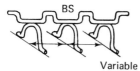

Beam Bolster

Heavy Beam Bolster

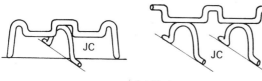

Variable

Heavy Beam Spacers

High Chairs

Joist Chairs

Fig. 11-16   Accessories for reinforcing bar.

Fig. 11-17   Designation of reinforcing wire mesh.

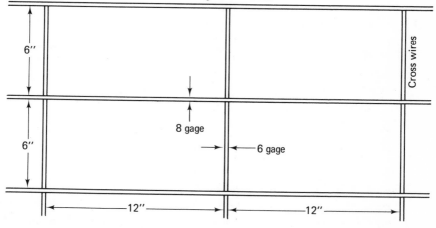

and a preferred direction and can be laid in one direction only; two-way mesh has square openings and can be laid lengthwise or crosswise. When the mesh is lapped end to end, the overlap should be one mesh opening. Wire spacing varies from 2 to 16 in. in both directions. Such mesh is designated usually by four numbers that give the weave and wire sizes; thus $6 \times 12 \times \frac{8}{4}$. The first number is the spacing of the longitudinal wires (Fig. 11.17), the second number is the spacing of the transverse wire. The two numbers $\frac{8}{4}$ are the gages: the longitudinal wires are No. 8 in size and the transverse wires are no. 4.

Wire gages correspond closely, but not exactly, to gage numbers for sheet steel, for example:

| Wire Gage | Diameter (in.) |
|-----------|----------------|
| 10 | 0.1350 |
| 8 | 0.1620 |
| 6 | 0.1920 |
| 4 | 0.2253 |
| $\frac{1}{4}$ in. | 0.2500 |
| 2 | 0.2625 |
| 1 | 0.2830 |
| 0 | 0.3065 |

Removable or permanent steel forms for concrete joist construction are made in 16-gage or heavier smooth steel sheet for removable forms, and 24-gage corrugated steel for permanent forms. Standard widths are 20 and 30 in. and the standard length is 3 ft. Available form heights range from 6 to 14 in. in increments of 2 in.

## 11.16 expanded metal mesh

*Expanded metal mesh* is made from metal sheet by slitting and expanding the sheet into diamond-shaped openings. This material has a very wide range of uses: for metal walkways and gratings, partitions, screens, etc. Such materials as stucco, plaster, troweled insulation, and concrete are readily attached to it because of its perforations, and special types of this mesh in lighter sheet gages are used in the plastering and insulating trades.

The usual method of designating the size of expanded metal mesh uses two numbers, the first number for the size of the mesh and the second for the gage of the sheet. Thus $1\frac{1}{2}$–13 is a $1\frac{1}{2}$-in. mesh of 13-gage material. The diamond shape of the mesh opening has a length of somewhat more than twice the width. The mesh size refers to the shorter dimension, in this example

$1\frac{1}{2}$ in. wide (which means approximately 3 in. long). This example is a heavy mesh suitable for a walkway or platform.

Figure 9.1 shows a standard expanded mesh. Mesh is also available flattened.

## 11.17    steel roofing, siding, decking, and partition studs

The uses of steel sheet, plain or galvanized with zinc, is widespread for roofing, siding, and decking applications, and is most familiar in the many types of prefabricated steel buildings. For such applications the steel is corrugated, ribbed, or cold-formed to provide stiffness in the thin gages used. Such formed sheets are supplied in length increments of 1 ft up to 12-ft lengths.

A comparison of *steel roofing* and *siding* in corrugated sheet is given in Fig. 11.18. The siding sheet turns both edges of the corrugated sheet in the same direction, while the roofing sheet has one edge corrugated up and the other down. The net width of both types of sheets is 24 in., as shown in the figure.

Fig. 11-18    Corrugated steel siding and roofing sheets.

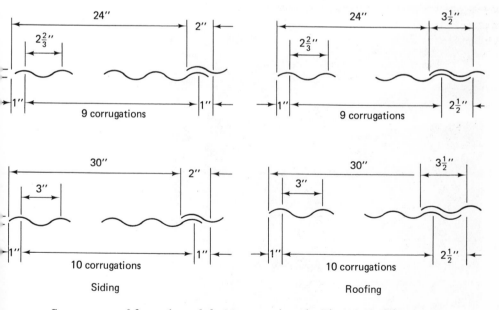

Siding

Roofing

Some types of formed *steel decking* are given in Fig. 11.19. These may be as light as 22 gage. Both the open face and cellular type of deck are shown in the figure.

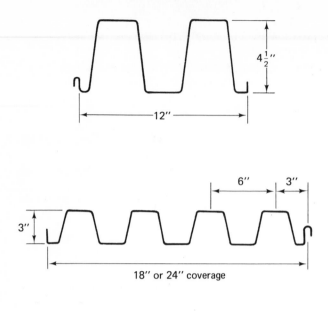

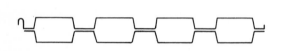

Fig. 11-19   Open and cellular types of steel decking.

Fig. 11-20   Installation of steel studs for a partition.

Steel decking is often used as a base and bottom form for a light concrete slab or floor, a built-up roof, a wood floor, or a sprayed polyurethane roof.

Figure 11.20 shows the erection of *steel studs* used to replace wood studs in a non-load-bearing partition. The studs are light-gage formed steel channels that snap into ceiling and floor channel tracks and are then faced with wallboard. Stud widths are $1\frac{5}{8}$, $2\frac{1}{2}$, or $3\frac{5}{8}$. The steel studs are perforated with pass-through holes for building services such as pipe and electrical conduit.

### 11.18   open-web steel joists

*Open-web steel joists* of the type shown in Fig. 11.22 are lightweight welded Warren trusses, made in both shortspan and longspan types. Shortspan joists are produced in lengths from about 4 to 48 ft, and depth from 8 to 24 in. Longspan joists are heavier, in lengths from 22 to 96 ft and depths 18 to 48 in.

Fig. 11-21   Open-web steel-joist construction.

Open-web steel joists provide direct support by the top chord for a roof deck or a floor. The top and bottom chord of the joist are made of two angles or other symmetrical section; an unsymmetrical section is avoided because it produces an eccentric stress in the chord and a tendency to rotate the chord. The web member is a continuous solid bar bent to the required configuration.

Attachments for mechanical and electrical services to the joist are preferably made with the use of clamping devices or U-bolt connectors, rather than drilling or cutting the joist. The casual cutting and drilling of these members is a dangerous practice.

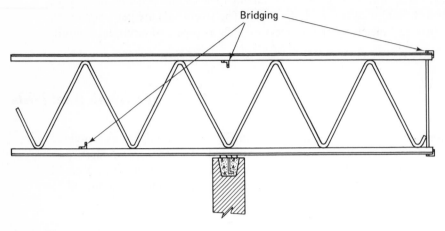

Fig. 11-22   Open-web steel joists.

## 11.19   welding of steel

Field welding operations on steelwork may include *arc welding, oxyacetylene welding,* or *oxyacetylene cutting.* Very light gage steel such as roof decking or any material lighter than 16 gage must be gas-welded. Arc welding is suited to 16 gage or heavier work. The oxyacetylene flame has a temperature of about 6000°F, while the temperature of the arc exceeds 10,000°F.

The welding of metals causes them to shrink or upset. If two pieces of steel exactly 12.000 in. long are butt-welded together side by side, the length of the weld will be slightly less than 12 in. (Fig. 11.23). Because of this shrinkage the pieces of steel will warp, and this warpage will be greater for

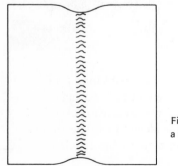

Fig. 11-23   Shrinkage produced by a butt weld.

lighter gages. Again, if the two pieces are to be fillet-welded together to make a 90-degree angle, if they are initially supported at 90 degrees to each other before welding, after welding the angle between them will be less than 90 degrees (Fig. 11.24).

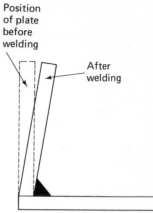

Fig. 11-24   Shrinkage produced by a fillet weld.

Arc-welding rods are available to deposit weld metal in various ultimate tensile strengths: 60,000 psi, 70,000 psi, and higher. Arc welding rods are designated by a four-digit number, the first two digits giving the tensile strength of the deposited metal. Thus 6010, 6014, 6024, etc., rods deposit weld metal with a tensile strength of 60,000 psi minimum, while 7018 rods deposit metal with a 70,000-psi minimum tensile strength. The final two

Fig. 11-25   Cross sections of weld fusion zones at various arc voltages. The increased depth of penetration at higher arc voltages should be noted.

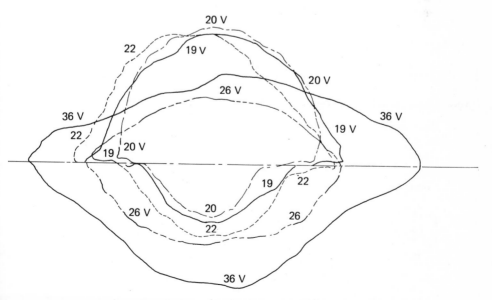

digits give information about other characteristics of the welding rod. If the third digit is 1, the rod is an all-position rod for welding horizontally, vertically, or overhead. If the third digit is a 2, only horizontal and flat welding seams may be deposited, since the rod metal does not freeze fast enough for vertical or overhead welding. Handbooks should be consulted for the more complex information conveyed by the fourth digit in the rod designation.

Some of the more useful welding rods are the following:

1. *6010:* deep penetration rods for butt welding of pipe, heavier gages of sheet, plate, and structural shapes. Requires a direct-current electrical power source.
2. *6011:* same as 6010, but will operate with either direct current or alternating current.
3. *6012:* medium penetration. A good rod to use where the fit between the plates to be welded is poor.
4. *6013:* shallow penetration. For thin-gage steel, such as 16 gage.
5. *6024:* suited to horizontal fillet welds. High deposition rate.
6. *6027:* similar to 6024.
7. *7018:* high-quality rod for ensuring crack-free welding of steel in cold weather.

Fig. 11-26   High-strength steel bolts for steel erection.

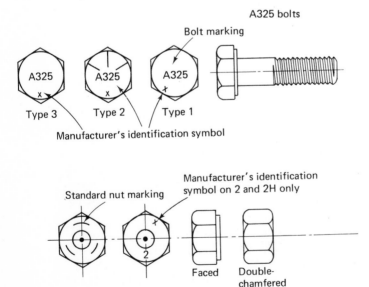

A325 bolts

Bolt marking

A325        A325        A325

Type 3      Type 2      Type 1

Manufacturer's identification symbol

Manufacturer's identification symbol on 2 and 2H only

Standard nut marking

Faced    Double-chamfered

*Rough hardware* includes all the miscellaneous small hardware items that form part of the structure of the building, including foundation, frame, and roof. Rough hardware therefore includes bolts, anchors, nails, ties, joint hangers, etc. Most of this hardware is concealed. *Finish hardware* comprises those items required in the finishing of a building, such as locks, door hinges, and window fasteners. Rough hardware is usually of steel, while finish hardware may be made of aluminum, bronze, stainless steel, or other decorative metals.

The high-strength bolts, nuts, and washers used in steel erection are made of high-tensile steel in accord with ASTM A325-58T. The dimensions and head markings for a high-strength bolt are shown in Fig. 11.26. In determining the bolt length required, add to the thickness of the bolted material the amounts given in the following table, to allow for a nut and two flat washers:

| Bolt Size (in.) | Add (in.) |
| :---: | :---: |
| $\frac{1}{2}$ | 1 |
| $\frac{5}{8}$ | $1\frac{1}{8}$ |
| $\frac{3}{4}$ | $1\frac{1}{4}$ |
| $\frac{7}{8}$ | $1\frac{1}{2}$ |
| 1 | $1\frac{5}{8}$ |
| $1\frac{1}{8}$ | $1\frac{3}{4}$ |
| $1\frac{1}{4}$ | $1\frac{7}{8}$ |

For bolting to flanges of I beams, channels, and angles, bevelled washers are required, shown in Fig. 11.27.

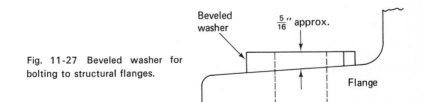

Fig. 11-27 Beveled washer for bolting to structural flanges.

Other types of fastening devices used in building construction include rivets, lag bolts, wood screws, toggle bolts, expansion anchors for holding bolts in holes drilled in concrete, and various styles of nails.

Some of the many kinds of nails are illustrated in Fig. 11.28. The small-head nails are used for light nailing and finishing work, while nails for heavy fastening require large heads. The nail styles with annular, helical, or other deformed shanks have greatly increased holding power but cannot be remov-

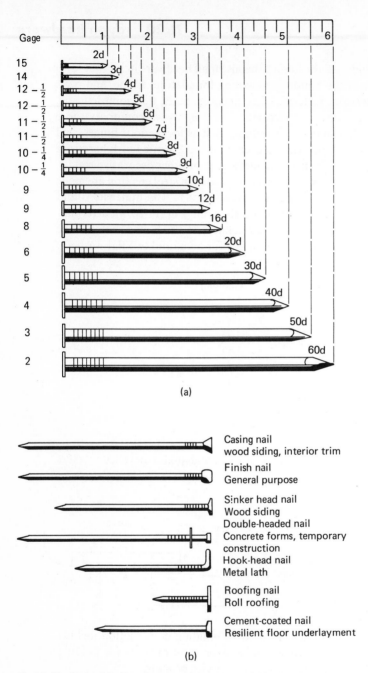

Gage

15
14
12 - $\frac{1}{2}$
12 - $\frac{1}{2}$
11 - $\frac{1}{2}$
11 - $\frac{1}{2}$
10 - $\frac{1}{4}$
10 - $\frac{1}{4}$
9
9
8
6
5
4
3
2

2d
3d
4d
5d
6d
7d
8d
9d
10d
12d
16d
20d
30d
40d
50d
60d

(a)

Casing nail
wood siding, interior trim

Finish nail
General purpose

Sinker head nail
Wood siding

Double-headed nail
Concrete forms, temporary
construction

Hook-head nail
Metal lath

Roofing nail
Roll roofing

Cement-coated nail
Resilient floor underlayment

(b)

Fig. 11-28   Nails: (a) sizes of common nails; (the gage sizes are
wire gages) ; (b) various types of nails.

ed without destroying the material that is nailed. Steel nails must not be exposed to moisture, as they will corrode and stain adjacent materials or even rot wood. Galvanized nails such as roofing nails can be exposed to moisture.

Various types of hangers are required to make structural connections in building frames. Some of the many types of hangers are illustrated in Fig. 11.29. These hangers include wood-to-wood hangers, wood-to-steel joist hangers, and steel-to-steel beam hangers.

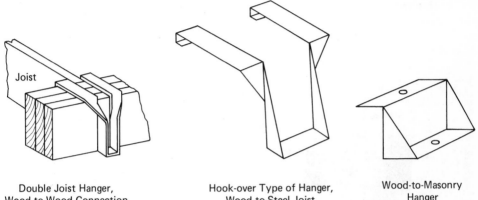

Double Joist Hanger, Wood-to-Wood Connection

Hook-over Type of Hanger, Wood-to-Steel Joist

Wood-to-Masonry Hanger

Fig. 11-29   Types of sheet-steel hanger connections.

## 11.21   hardness of metals

The hardness of materials must often be known, because hardness is the clue to other properties of materials. A harder steel is, for example, a stronger steel, more wear-resistant, and probably less ductile. Hardness tests must also be made on metals to check the success or failure of a welding operation or a heat treatment. A welding operation that produced hard areas in a steel building frame would be viewed as dangerous, since the hardened areas would be brittle.

There are a considerable number of hardness tests in use, chiefly because a single hardness method cannot span the whole range of hardness from the soft plastics such as polyethylene to the hard ceramic materials such as quartz. However, almost all hardness methods measure indentation resistance to some kind of indentor. The indentor is pressed into the specimen; a small indentation indicates a high hardness, a large indentation indicates a low hardness. The Moh's scale of hardness discussed in Section 4.2 was an exception; it is a scratch test for hardness.

Through experience everyone acquires the notion that hardness is related to strength in tension and compression. Metals are both hard and strong; the plastics usually are both soft and of low strength. Harder metals are stronger metals.

## 11.22   brinell hardness test

A *Brinell hardness tester* is shown in Fig. 11.30. A hardened steel ball 10 mm in diameter is forced into the test part at a load of 3000 kg for steels or 500 kg for nonferrous metals such as brass or aluminum. The Brinell test should not be made on very hard steels, since these may deform the penetrating ball. Neither should this test be made on thin material, such as plate less than $\frac{1}{4}$ in. thick, such that the underside of the plate shows bulging due to penetration from the ball.

Fig. 11-30   Brinell hardness tester.

The Brinell hardness number (BHN) is found from the diameter of the impression. This diameter is read in millimeters on a special Brinell low-power microscope with a built-in scale. The BHN is found from a table of millimeter values and BHN numbers. The BHN is actually related to the diameter of the impression by the formula

$$\text{BHN} = \frac{P}{(\pi D/2)(D - \sqrt{D^2 - d^2})}$$

where

$P =$ Brinell load, kilograms
$D =$ diameter of indentor ball
$d =$ diameter of impression in the specimen

The denominator in the formula is the surface area of the impression, so the Brinell hardness number is actually the indenting force per square millimeter of the impression.

The ultimate tensile strength of a steel can be closely approximated by multiplying the BHN by 500.

## 11.23   rockwell hardness test

The Brinell test is neither as quick nor as convenient to perform as the *Rockwell test*. It leaves a rather large impression, which is usually damaging to the part tested. The Rockwell impression is small enough that sometimes such an impression may not harm the part tested.

Fig. 11-31   Rockwell hardness tester.

The Rockwell tester (Fig. 11.31) uses a hardness dial graduated in two scales, Rockwell B and Rockwell C, abbreviated $R_b$ and $R_c$. Rockwell B tests are performed on soft steels, including structural steels; the Rockwell C test must be used with hardened steels. The $R_b$ test uses a $\frac{1}{16}$-in.-diameter hardened ball with a 100-kg load. A 10-kg load is first applied by the testing machine. The operator then turns the dial to the SET mark and applies the major load of 100 kg. The major load is then removed and the hardness read on the scale. The $R_c$ test uses a conical diamond indentor, called a Brale, with a major load of 150 kg. The procedure is the same as for an $R_b$ test. The Rockwell tester actually measures the further indentation of the major load beyond the indentation of the minor load, but the scale is graduated in Rockwell numbers instead of depth of penetration.

Any movement of the specimen under test will produce a false reading. Very often the first Rockwell reading on a specimen is untrustworthy and is disregarded.

Some significant Rockwell hardness numbers are the following:

1. Mild steel has readings in the range $R_b$ 60–80. These hardnesses are less than $R_c$ zero.
2. The maximum hardness obtainable in any steel is $R_c$ 65–67. Only high-carbon steels will give such a high reading.

## 11.24 aluminum

*Aluminum* and its alloys are about one-third the weight of steel. This light weight is one of the chief advantages of aluminum for construction. Besides its light weight, aluminum has an attractive silver-white appearance and is resistant to rainwater and atmospheric corrosion. It is protected from corrosion by a thin film of its own oxide, alumina. Like steel, this metal is available in a range of alloys that provide a range of properties. It has two significant disadvantages, however: a high coefficient of thermal expansion

Fig. 11-32 A geodesic dome constructed of aluminum alloy 6061 tubing, at the headquarters of the American Society for Metals, Cleveland.

(twice that of steel) and a modulus of elasticity one-third that of steel. This lesser stiffness must be allowed for in design, so that if aluminum is substituted for steel, the aluminum sections will be somewhat larger than would be the case for steel. The aluminum frame therefore would not be one-third as heavy, as the relative weights of the metals suggest, but half as heavy.

Pure aluminum is a very soft and weak material, with an ultimate tensile strength of about 11,000 psi. Like pure iron, pure aluminum is not used in building construction. Aluminum can be alloyed to provide strengths exceeding that of mild steel. The very strong alloys, however, have very limited ductility and limited resistance to atmospheric corrosion, and are not usually employed in construction.

The alloys of aluminum used in construction applications fall generally into two groups. Flat and corrugated aluminum sheet is used for roof flashings and wall panels, uses that require ductility and corrosion resistance. For best corrosion resistance a relatively pure aluminum alloy is preferred. The usual alloy for such uses is 3003, which is alloyed with a small amount of manganese. The other type of application is the many uses of aluminum in special shapes: window moldings, building trim, channels, and H beams. Some of these shapes are shown in Fig. 11.33. These shapes are produced from alloys of medium strength, in the range 40,000 psi ultimate tensile strength, and are almost always magnesium or magnesium–silicon alloys of aluminum.

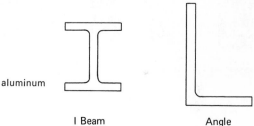

Fig. 11-33 Extruded aluminum structural shapes.

I Beam          Angle

Steel shapes are rolled from ingots on large and expensive roll stands. Aluminum shapes are extruded from small round billets through extrusion dies. Extrusion dies are relatively inexpensive, and therefore it is often economical to design a special aluminum shape for structural or trim purposes. The die charge can be paid off in a few hundred feet of extrusions. Such low costs account for the infinite variety of aluminum moldings that one encounters in building construction (Fig. 11.34).

Fig. 11-34 Extruded trim in aluminum.

Door Post          Threshold          Coping

Steel angles and other steel structural shapes require a "draft" or taper in the flanges in order to be successfully rolled. Extruded shapes do not require draft, which therefore is not a feature of aluminum angles, channels, and H beams (see Fig. 11.33).

Architectural aluminum, such as the shapes shown in the building entrance of Fig. 11.35, are usually *anodized.* Anodizing is an electrochemical bath process which thickens the protective natural oxide coating on the aluminum, thus increasing the corrosion resistance and hardness and improving appearance and cleanliness. The anodized surface can also hold dyes or enamels for coloring effects. Unanodized aluminum does not hold paint well, unless an undercoat of zinc chromate is used.

Fig. 11-35 Frames and doors made of extruded aluminum shapes.

Aluminum has unusually high reflectivity for light and heat, although it reflects heat less well than light. Aluminum foil is used as a reflective thermal insulation, serving also as a vapor barrier.

Hardware for the attachment of aluminum should be of the same aluminum alloy or of stainless steel if galvanic corrosion is to be prevented. There can, of course, be no galvanic corrosion except in the presence of water or condensing water vapor. In the presence of water and another metal (except stainless steels of the 300 series) aluminum will be rapidly corroded. Water discharged from copper piping that has splashed over aluminum has corroded the aluminum.

Aluminum can be attacked by damp concrete or plaster during setting of these materials. Such corrosion is prevented by a coat of bituminous paint. Even oak and western red cedar can corrode aluminum, since there are acids in these woods.

*Copper* is used in buildings for its corrosion resistance and its attractive colors. By suitable alloying, metallic colors from silver white through yellow, red, and brown are possible with copper. However, through long exposure to atmosphere, copper will weather to a characteristic green color, which is often the color that the architect ultimately intends. The color changes initially to a dark brown, but over five years becomes the characteristic green "patina." The change of color is prevented with suitable lacquer coatings.

Fig. 11-36   Sheet-copper roof.

Copper is used in sheet form for construction, as roof flashings, roofing sheets (Fig. 11.36), or interior and exterior ornamental work. Special laminated types of copper flashing sheet are available, such as a copper sheet bonded with a layer of asphalt to a heavy kraft (brown) paper, or copper bonded with asphalt to a lead and a kraft paper backing. The backing materials protect the concealed surface of the copper sheet from corrosion. Examples of flashing practice are shown in Fig. 11.37.

Fasteners for copper should be copper or stainless steel if galvanic corrosion is to be avoided. Like aluminum, copper has a greater thermal expansion than steel.

Alloys of copper with zinc are called *brasses*. Alloys of copper with metals other than zinc are called *bronzes*. The bronzes have much better atmospheric and corrosion resistance than the brasses, hence their use for plaques and ornamental work. The number of brasses and bronzes is very

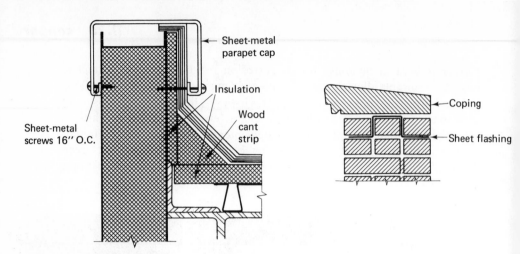

Fig. 11-37   Sheet-metal flashing.

large, and their names and characteristics are confusing (low brass, red brass, commercial bronze, muntz metal, and many others). A brief and intelligible discussion of these materials cannot be undertaken here.

## 11.26  zinc and lead

*Zinc* finds extensive employment as the coating on galvanized steel, serving to protect steels from atmospheric corrosion. If galvanized steel is to be painted, it requires a suitable primer to ensure bond of the paint to the zinc coating, although, as discussed below, certain types of zinc coatings are specially prepared to receive paint.

Zinc can oxidize rapidly to white zinc oxide if the zinc surface is enclosed and if it is kept in a wet condition. This type of corrosion, called *white rusting*, does not occur on surfaces exposed to the weather.

Seven types of galvanized steel sheet are produced. These seven types fall into two broad classes: hot-dipped (in molten zinc) and electrocoated by electroplating methods.

### HOT-DIPPED GALVANIZED STEEL SHEET

*No. 1:*  hot-dipped galvanized steel, spangled, in various coating thicknesses. This type has the familiar spangled appearance of galvanized steel. It is not suited to receive standard paints.

*No. 2:*  hot-dipped and heat-treated to produce an alloyed zinc–iron coating, in various thicknesses. This type is often named "Galvannealed." It has a dull gray color without spangle and is intended for painting.

*No. 3:* hot-dipped and phosphatized to receive paint.

*No. 4:* hot-dipped, with a lighter zinc coating on one side, also termed "Differentially Coated." This type has the regular spangled appearance.

## ELECTROLYTIC ZINC-COATED STEEL SHEET

*No. 5:* electrolytic zinc-coated, with 0.1–0.2 oz of zinc per square foot of sheet (total on both sides).

*No. 6:* electrolytic zinc-coated and phosphatized to receive paint.

*No. 7:* electrolytic flash zinc-coated and phosphatized for paint.

The weight of the zinc coating (total on both sides) is designated by a letter, G for Galvanized and A for Alloyed, and a number, which represents the ounces per square foot. Thus G235 has 2.35 oz/ft$^2$ and A60 has 0.60 oz/ft$^2$. The useful life of the galvanized sheet is proportional to the coating thickness. The heavier coatings such as G235 are preferred for exterior applications.

Electrolytic zinc-coated steel receives a zinc coating about one-tenth as thick as the hot-dipped grades and is less used in construction.

*Lead* is a soft, pliable, and heavy metal, with remarkable corrosion resistance. It is used in sheet form for a limited number of building applications, including roofing and flashing. A square foot of lead $\frac{1}{16}$ in. thick weighs 1 lb.

Lead is an excellent surface for reducing the transmission of sound through walls, or for reducing the transmission of gamma radiation.

## PERTINENT ASTM SPECIFICATIONS

E8-69 (Part 6 and 31) Tension Testing of Metallic Materials

E9-67 (Part 31) Compression Testing of Metallic Materials

E10-66 (Part 31) Brinell Hardness of Metallic Materials

E18-67 (Part 31) Rockwell Hardness of Metallic Materials

E140-67 (Part 31) Hardness Conversion Tables for Metals

E16-64 (Part 31) Free Bend Test for Ductility of Welds

E190-64 (Part 31) Guided Bend Test for Ductility in Welds

E23-66 (Part 31) Notched Bar Impact Testing of Metallic Materials

E208-66T (Part 31) Drop-Weight Test for Transition Temperature in Steels

E206-66 (Part 31) Definitions Relating to Fatigue Testing

D2478-66T (Part 16) Terms Relating to Nails

D1761-64 (Part 16) Testing Metal Fasteners in Wood

A307-68 (Part 4) Low Carbon Steel Threaded Standard Fasteners

A325-71a (Part 4) High Strength Bolts for Structural Joints

A370-71b (Part 3, 4, or 31) Mechanical Testing of Steel Products
A6-71 (Part 4) General Requirements for Rolled Steel Shapes
A36-70a (Part 4) Structural Steel
A82-70 (Part 4) Cold-Drawn Steel Wire for Concrete Reinforcement
A496-70 (Part 4) Deformed Steel Wire for Concrete Reinforcement
A615-68 (Part 4) Deformed Billet-Steel Bars for Concrete Reinforcement
A616-68 (Part 4) Rail Steel Deformed Bars for Concrete Reinforcement
A617-68 (Part 4) Axle Steel Deformed Bars for Concrete Reinforcement
A167-70 (Part 3) Stainless Chromium–Nickel Plate, Sheet, and Strip
B209-72a (Part 6) Aluminum Alloy Sheet and Plate
B221-72 (Part 6) Aluminum Alloy Extruded Shapes

## QUESTIONS

1  Define steel.

2  What is the difference between a plain carbon and an alloy steel?

3  What is the purpose of the coke added to the blast furnace charge?

4  Why are steel products not rolled from the hot metal tapped from the blast furnace?

5  What is the method used to reduce the carbon in an oxygen converter?

6  What is the reason for inoculating steels with small amounts of (a) manganese; (b) silicon; (c) aluminum?

7  What methods are available for lowering the transition temperature for brittle fracture?

8  What elements when added to steel will lower the transition temperature for brittle fracture?

9  What is the influence of carbon on this transition temperature?

10  Differentiate between a killed and a rimmed steel.

11  All arc welding rods are very low in carbon. Why?

12  State the approximate carbon content of low-carbon, medium-carbon, and high-carbon steels, and state the general applications for each of these broad types of steel.

13  Structural steels are low-carbon steels. Why?

14  State which of the following might serve as structural steel: (a) 0.80% C, 0.30% Si, 0.60% Mn; (b) 0.13% C, 9% Ni; (c) 0.2% C, 0.30% Si, 0.90% Ni, 0.60% Cu; (d) 0.38% C, 5% Cr, 1% Si. State your reasons in each case.

15  What is the approximate alloy composition of a 300 series stainless steel?

16  What simple method would differentiate between a 300 and a 400 stainless steel? If you don't know, ask a dealer in scrap steel how he differentiates between them.

**17** The wall of a 100,000-gal water-storage tank measures 30 ft in diameter by 26 ft high, and is made of $\frac{1}{4}$-in. steel plate. What is the weight of the wall of the tank?

**18** Name the gage of steel sheet that has each of the following approximate thicknesses: (a) $\frac{1}{32}$ in.; (b) $\frac{1}{20}$ in.; (c) $\frac{1}{8}$ in.; (d) $\frac{1}{16}$ in.

**19** What would you state to be the difference between a pipe and a tube?

**20** State the three grades of reinforcing bar.

**21** What is the diameter of No. 3 and No. 6 rebar?

**22** Why would you not let subtrades casually cut and modify an open-web steel joist?

**23** Select a welding rod for the following purposes: (a) the welding of 16-gage steel sheet; (b) a butt weld in $\frac{1}{4}$-in. steel plate; (c) a horizontal fillet weld joining $\frac{1}{4}$-in. steel plate; (d) a butt weld in the $\frac{7}{8}$-in. steel plate of a cement kiln, where reliability of the weld is very critical.

**24** Differentiate between rough and finish hardware.

**25** Why are alloys of aluminum used instead of pure aluminum?

**26** (a) Why are aluminum shapes extruded instead of rolled? (b) Why are steel shapes rolled instead of extruded?

**27** Give some of the reasons for anodizing aluminum.

**28** (a) Differentiate between a brass and a bronze. (b) Which of these two types of copper alloy will have the better corrosion resistance?

Wood is one of the natural organic polymer materials discussed in Chapter 14. Wood is actually a cellular material. In recent years a great many cellular plastics and rubbers have appeared: by a wide margin wood is the strongest of these.

Although most familiar as lumber and timber for structural purposes, wood is also used in a range of other forms. These include plywood, glued laminated timbers, particle board, wood fiber for gypsum wall panels, hardboard, and other building products. Wood is also the raw material for paper and paperboard, a considerable number of plastics, and nitrocellulose for explosives. Although the use of wood in its natural structure appears to be declining, there is an increasing use of wood products in which the wood is modified chemically or mechanically, as in hardboard or particle board. Currently, foamed polyvinyl chloride is substituting for wood in such products as shelving, house siding, moldings, and fence posts. This foamed plastic bears a rather remarkable likeness to wood.

## 12.1 structure of wood

A tree grows by the addition of new wood to the outer layer of the tree, the *cambium* layer immediately underneath the bark of the tree. This annual

# WOOD
# PRODUCTS

# 12

Fig. 12-1 Growth rings in Douglas Fir. At the bottom-right-hand corner, the rings indicate 10 dry years followed by 5 rainy years.

growth can be seen in the concentric series of growth rings (Fig. 12.1). Each year adds one more growth ring to the diameter of the tree, at least in temperate zones under normal conditions. It is preferable to call these rings *growth rings* rather than annual rings, since there are cases in which more than one ring may be formed in a single year, although this is unlikely in temperate or cold climates. In the tropics the growth rings correspond with the alternating wet and dry periods of the seasons. The cambium, which produces both wood and bark, is visible only under the microscope.

The growth rings are easily distinguished from one another because of color differences between *earlywood* and *latewood*, also known as *springwood* and *summerwood*, respectively. These two types of annual growth may also differ in hardness and density, especially in some softwoods, such as Douglas fir. Latewood (summerwood) is darker in color, heavier, and harder, because later in the season the cells of the tree grow more slowly. Springwood contains cells with larger cavities and thinner wall due to the more rapid rate of growth in the earlier part of the growing season. The distinction between earlywood and latewood is more difficult in the case of tropical hardwoods.

As the tree continues to add annual rings, the older, more central rings gradually cease to contribute to the physiological processes of the tree, such as the movement and storage of food chemicals, and provide only mechanical support to the tree. This functional change produces what is called *heartwood* in the center area of the tree (Fig. 12.2). The heartwood portion is usually darker in color, drier, and harder than the layers that are

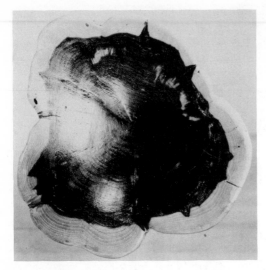

Fig. 12-2 Heartwood and Sapwood in a Red Cedar.

still physiologically active. The outer, living part of the tree is the *sapwood*. The diameter of the heartwood will decrease toward the top of the tree trunk, since the growth rings are a series of concentric cones.

All woods have radial marks called *medullary rays*, extending from the bark toward the central pith of the tree trunk. These rays are quite conspicuous in some hardwoods, such as oak and birch, but may be difficult to find in softwoods.

Chemically, wood is a compound polymer composed of three principal polymeric constituents. The nature of polymeric substances is discussed in Chapter 14; for the moment, a polymer may be considered to be a chemical that consists of a very long chain of atoms. The wood of a tree consists of long tubular fibers. These fibers are composed of two polymeric carbohydrates called *cellulose* and *hemicellulose*. Both are complex glucose compounds. Glucose is a sugar, which explains why fungi and insects, and even animals, can use wood as a food. These wood fibers are bound together with a third polymer, called *lignin*. These fiber bundles run the length of the tree or the branch and carry food products from the roots to the leaves. Based on an oven-dry condition, the constitution of woods is the following:

|               | Hardwoods (%) | Softwoods (%) |
|---------------|:-------------:|:-------------:|
| cellulose     | 40–45         | 40–45         |
| hemicellulose | 15–35         | 20            |
| lignin        | 17–25         | 25–35         |

Small amounts of other substances are also present in woods.

It is cellulose that provides the strength in axial tension, toughness, and elasticity of wood. The long-chain molecules of cellulose are in bundles that run helically to form hollow needle-shaped cells or fibers. These fibers range in length from 1 to 3 mm. The fibers have thicker walls in hardwoods than in softwoods. The long-chain structure of cellulose makes it able to form fibers similar to other vegetable fibers, such as cotton. The chemical structure of cellulose is given in Fig. 12.3 as $(C_6H_{10}O_5)$ repeated $n$ times, where $n$ is a very large number between 8000 and 10,000.

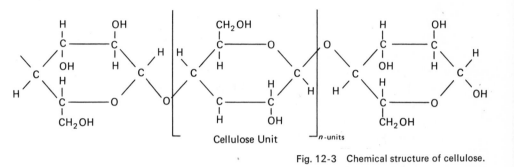

Fig. 12-3   Chemical structure of cellulose.

Associated with the cellulose bundles are a group of similar polymeric substances, the hemicelluloses. These are chemically similar to cellulose. The chain length is shorter, about 150 units, so the hemicelluloses are not fibrous but are gelatins.

Lignin bonds the individual fibers together into wood, giving the wood its compressive strength. The chemical structure of lignin is not completely known. The manufacture of paper requires the removal of most of the lignin in order to free the individual wood fibers. It is rather interesting that lignin is found only in association with cellulose, although cellulose may be found independently, as in cotton.

## 12.2   hardwoods and softwoods

The terms "hardwood" and "softwood" do not differentiate between woods that are hard or soft. The summerwood of Douglas fir, a softwood, is extremely hard, while balsa wood, a hardwood, is the softest of all woods. The *hardwoods* are the deciduous or broad-leaved trees which shed their leaves in the fall in cooler climates. *Softwoods* are the conifers that bear needles rather than leaves and produce their seeds in cones. Tamarack, which is a softwood, nevertheless sheds its needles in the fall. Softwoods generally contain more lignin and less hemicellulose than most hardwoods. Both the lightest and the heaviest, the softest and the hardest, of all woods are tropical hardwoods. The lightest is balsa, weighing from 5 to 10 lb per ft³. The heaviest is lignum vitae or ironwood, weighing 75–80 lb per ft³.

Softwoods are found in many tropical areas, but are more characteristic of colder climates and higher altitudes, usually in large tracts of relatively few species. Hardwoods become less common as latitude increases. About 30 softwoods and 50 hardwoods are of commercial importance in the United States and Canada.

The individual character, color, and smell of the specific species is due to extractives in wood, mainly in the heartwood. These extractives are so called because they can be removed from the wood with solvents. The extractives may also make a timber resistant to decay or to insect and fungus attack. Despite their small quantity, about 1–2 per cent by weight, the extractives have a significant effect on the properties of the wood. These extractives explain the excellent durability of redwood, oaks, and cypress, and the durability of heartwood as compared to sapwood generally. Usually a dark color in wood indicates a durable timber. Many of these natural dyes in the wood, however, are destroyed or bleached by ultraviolet light, or can evaporate over a period of time.

## HARDWOODS

*Ash.* Heavy and hard, coarse-grained. Brown heartwood and whitish sapwood. Excellent impact resistance (used in baseball bats). Used for trim and furniture.

*Basswood.* Soft and light in color. Familiar in drafting boards.

*Beech.* Heavy and hard, white to brown. Used for trim and plywood.

*Birch.* Fine-grained and hard. Easy to finish, polish, and stain, with no pronounced pattern. Often stained to imitate other hardwoods. Used for furniture, plywood, finish, doors, and flooring. Not resistant to rot.

*Cherry.* Light color and close-grained. Reddish-brown heartwood and creamy sapwood. Used for trim and finish.

*Elm.* Heavy and hard, light brown.

*Mahogany.* Strong, but works easily and takes finishes well. Reddish brown. Used in furniture, plywood, doors, trim, and finish.

*Maple.* Soft pastel gray color. Easily worked and does not split or chip. Used for furniture, flooring, and interior trim.

*Oak.* Heavy and hard, open-grained. Takes a variety of finishes well. A poor wood for dimensional stability and difficult to work. Available in two common species, white oak and red oak, which are often similar in appearance; red oak has a coarser grain. Used for furniture, plywood, flooring, interior trim. Not recommended for exterior use because of warping with changing moisture content; red oak is also susceptible to rot.

*Poplar.* Light color, soft, uniform in color. Used for cheap grades of plywood.

*Walnut.* Hard, strong, dark, durable, and easy to work. Takes finishes well. Used in furniture, flooring, plywood, trim, and finish.

## SOFTWOODS

*Cedar, Western Red.* Soft, straight-grained, with pronounced tendency to split. Used in pencils because it is easily whittled. Heartwood is reddish, sapwood is limited in area in the tree and is a cream color. Resistant to rot. Used for shingles, siding, finish, fence posts.

*Cypress.* Close-grained. Bright yellow heartwood, light brown sapwood Used for trim and finish.

*Fir, Douglas.* A majestic tree that is impossible to replace because of the height and age of existing firs. Wood is hard, strong, and durable. Heartwood color ranges from reddish to yellow; while sapwood is whitish. Its uses are extensive in construction. Most familiar in plywood and structural uses. Latewood is dark, extremely hard and brittle, and tends to stand out in relief above the springwood.

*Hemlock.* Fine texture and straight grain, easily worked. Has all the uses of fir, for which it is often substituted.

*Pine.* Pines are soft, white, straight-grained, with little tendency to warp. White pine is soft, white, durable, and works easily even in complex shapes. Used for furniture and shelving. Sugar pine has a reddish heartwood; used for furniture and interior trim. Lodgepole pine is used as dimension lumber. Yellow pine or southern pine is hard and strong and more difficult to work. Heartwood is reddish. Used as a structural wood like fir, in beams and planks.

*Redwood.* Dull red color, easily worked, but, like cedar, not a strong wood. Used for interior and exterior trim and shingles.

*Spruce.* White spruce is soft and easily worked. Used as dimension lumber. Engelmann spruce is similar to white spruce in properties and uses. Sitka spruce is unusually strong and tough.

## 12.3 grain in wood

The word *grain* as applied to wood is indeterminate in its meaning. Conversationally it refers to the appearance or pattern of the wood on any of its cut surfaces. The figure or pattern of a wood is due to variations of ring growth and of color in the wood, together with the influence of knots. This pattern is preferably not referred to as "grain."

An *open-grained wood* such as oak has minute pores over its exposed surface. In standard wood-finishing methods, these pores are leveled with a coat of filler. A *close-grained wood* such as fir or pine has no such pores in its surface.

The grain of the wood also refers to the direction of the cellular or fibrous structure of the wood, which is the longitudinal direction. Timbers for structural use must be so cut that the grain runs parallel to the length of the timber; otherwise there is a marked reduction in strength.

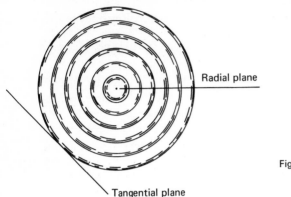

Radial plane

Fig. 12-4   Planes of a log.

Tangential plane

In a traverse section the log of wood looks like a series of concentric circles due to annual rings. The tangential plane is a plane at a tangent to these circles; the radial plane follows a diameter and passes through the center, as shown in Fig. 12.4. Planks sawed tangentially are termed flat-sawn, giving a *flat grain* (Fig. 12.5), while radially cut planks are termed quarter-sawed, giving an *edge grain*. If the annual rings are approximately at 45 degrees to the face of the plank, the condition is called *angle grain*. *Cross grain* refers to a plank whose fibers are not parallel to the long axis of the plank.

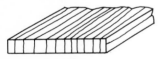

Edge Grain

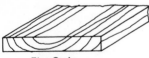

Flat Grain

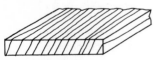

Fig. 12-5   Types of grain produced by sawing.

Angle Grain

Because wood is not a controlled factory product, it shows considerable variation in physical properties such as strength. All the strength properties, including shear strength, differ remarkably among the longitudinal, radial, and tangential directions. Strength is much higher in the longitudinal direction, about 10 times larger than the same strength parallel to the grain. Shear strength parallel to the grain is poor and undependable.

The allowable stresses of clear wood without defects depend upon the species, but range from 1000 to 2500 psi in bending parallel to the grain, for both hardwoods and softwoods. Allowable compressive stresses parallel to the grain are lower. All woods are considerably weaker in compression perpendicular to the grain than parallel to the grain. The modulus of elasticity may be as high as 1,600,000 for Douglas fir, southern pine, beech, birch, and maple, or as low as 800,000 for northern white cedar. Allowable stresses for hardwoods are not necessarily higher than those for structural softwoods. Allowable strength values must be reduced for defects in the wood.

The hardness of wood is proportional to its density. Wood flooring, meat-cutting blocks, and many other applications for wood require that the wood be abrasion-resistant. A hardwood is required for such purposes. The best wear resistance is given by an end-grain surface.

Softwoods have a thermal conductivity of close to 1 Btu/in. of thickness, which is the value of a moderately good insulator. Increased moisture content increases the thermal conductivity. The thermal expansion of wood is small and generally is ignored in design.

Certain woods contain weak acids that may corrode adjacent metals, although they do not corrode stainless steels. Oak particularly has damaged metals with which it has had contact. Steel fastenings will be attacked by red cedar, although brass fastenings are not. Wood itself, however, is remarkably corrosion-resistant.

Wood will be slowly weathered by ultraviolet radiation in sunlight, although the rate of weathering is extremely slow, about $\frac{1}{4}$ in. per century. The first effect of weathering is a bleaching of the color of the wood. The final effect of weathering is to reduce woods to a silver-gray color. The effect of water on an unprotected wood surface is to raise the grain and roughen the surface. With continued weathering the softer earlywood is eroded faster than the harder summerwood, producing an irregular surface. Almost all clear finishes such as varnishes are transparent to ultraviolet radiation. Hence the varnish film may not weather excessively, but ultraviolet action at the wood surface may cause a loss of adhesion of the varnish film to the wood.

**Specific gravity.**   The weights of materials are customarily given either in pounds per cubic foot or in specific gravity. *Specific gravity* is the ratio of the weight of 1 ft³ of a material to the weight of a cubic foot of water. In the metric system specific gravity is more simply the ratio of weight to volume. Since 1 ft³ of fresh water at standard temperature weights 62.4 lb, a wood with a specific gravity of 1.0 would weight 62.4 lb per ft³. Specific gravities of the more familiar woods range from 0.30 to 0.90. The differences in specific gravity between woods are accounted for by differences in the size of the wood cells and the thickness of the cell walls. The specific gravity of oven-dry cell wall material is 1.5.

Although wood will burn readily in thin sections, the fire resistance of heavy sections is exceeded only by some ceramic materials. The complete burning of, for example, a tree stump would be an almost endless operation. The surface of a fired wood chars to carbon, which is very slow burning, and the low thermal conductivity and the moisture content of wood delays it from heating to ignition temperature. If the fire resistance of wood must be improved, however, it may be impregnated with suitable chemicals such as ammonium salts or boric acid, or painted with fire-retardant paints.

## 12.5 moisture in wood

The affinity of cellulose for water is extremely high, and therefore wood readily takes up water. The wood changes its dimensions in response to moisture content, and these dimensional changes can be significantly large. The lighter woods can hold more moisture than heavy woods: balsa wood can in an extreme case hold 400 per cent moisture, while a heavy hardwood may be saturated at 30 per cent. Shrinkage due to moisture changes in wood may be as much as 0.1–0.2 per cent in the longitudinal direction. Shrinkage tangential to the growth rings may be as much as 5 per cent or more in changing from an oven-dry to a saturated condition. Radial shrinkage is about half the tangential shrinkage. Even the smaller changes in moisture content experienced by wood in service may cause problems, such as warping, the development of openings between boards, the loosening of nails, and the development of checks (cracks) in the wood. Checking is most frequently found at the ends of boards, since moisture can most easily penetrate the wood at the end grain. For least warping, edge grain is superior to flat grain (Fig. 12.6).

Moisture in wood is found in two conditions, *free moisture* and *combined moisture*. Free water is found outside the cell walls; combined water is present inside the cell walls, held between the long chains of the cellulose molecule. When wood dries, the free water evaporates first. If all the free water is driven out of the wood, but the combined water remains, the wood is at the "fiber saturation point," a condition giving about 30 per cent water content for all wood species. Evaporation of the combined water will cause the cellulose polymer chains to draw together and the wood to shrink. As noted previously, the shrinkage is predominantly tangential and to a lesser degree radial.

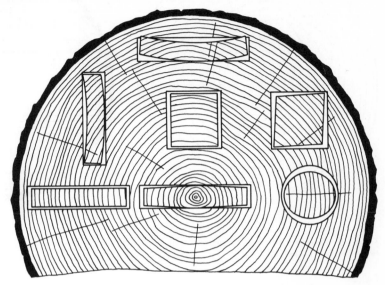

Fig. 12-6   Distortion and shrinkage as influenced by the method of cutting up the log.

The moisture content of wood is measured as

$$\text{moisture content } (\%) = \frac{\text{original wt moist wood} - \text{oven-dry wt}}{\text{oven-dry wt}} \times 100$$

Repeated oven-drying operations at 212°F are required until no more weight loss appears.

The amount of moisture movement and warping are both affected by the method of cutting the timber. Boards cut as in Fig. 12.7(a) are called flat or slash sawn; those like Fig. 12.7(b) are called rift or quarter-sawn. Since quarter-sawing is the ideal method of producing lumber, the ideal sawing method is that of Fig. 12.8(a). This method is not normally practical because a substantial amount of the log would be wasted. The method usually adopted for sawing up a log is that of Fig. 12.8(b), which is a kind of compromise between best quality and least waste.

Fig. 12-7   Flat-sawed and rift-sawed (quarter-sawed) boards.

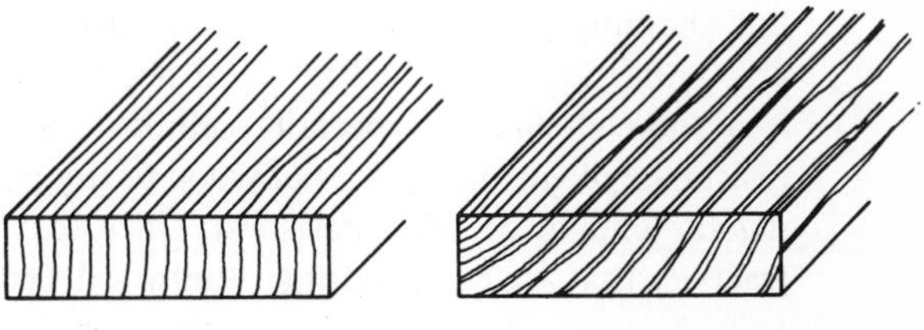

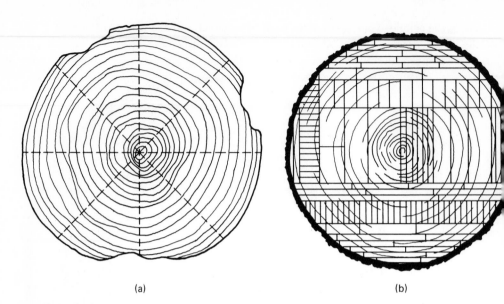

(a)                                                    (b)

Fig. 12-8   Sawing methods: (a) quarter-sawing of a log; (b)
typical sawing pattern for a log.

After sawing, the lumber is dried, either by storing in piles (air-drying) or in a drying kiln (kiln-drying). Ideally, the dried lumber should have the moisture content which will be in equilibrium with the relative humidity that it will encounter in service, although this condition is hardly obtainable in practice. The moisture content is reduced to about 15 per cent of the oven-dry weight of the wood, or less in kiln-drying.

## 12.6   wood decay

Fungi are the cause of decay in standing trees and wood products. This fungal decay is a disadvantage of wood as a construction material. Once the fungus has been established, its attack spreads. The spread takes place most rapidly along the axis of the tree.

Fungal attack requires four essentials, and without all four there is no attack: food, moisture, air, and temperature. If any of these factors is missing, the fungi will not die, but will lie dormant and revive when more favorable conditions return. The wood itself is food for the fungus, since wood is a food chemical, especially cellulose, a chemical similar to sugar. The resistance of various species of wood to fungal attack varies. In general darker woods tend to be more resistant, since they contain larger amounts of extractives that are toxic to fungi. Redwood, cedar, oak, mahogany, and walnut are among the more resistant species. Heavier woods are not necessarily more resistant.

Moisture and air conditions in wood are related: the space not occupied by water in the wood is occupied by air. In general, wood must contain at least 20 per cent moisture, based on oven-dry weight, for fungal attack. Very high moisture content will inhibit attack: wood, for example, does not rot when completely immersed in water. The necessity for air is apparent in the behavior of a rotted fence post: the rot is at and near ground level and the buried part of the fence post is sound.

Except for certain more unusual types, all rotting of wood falls into two chief types: brown rot (dry rot) and white rot. The dry rot condition (Fig. 12.9) splits the wood longitudinally and across the grain into cubes that are not usually larger than about $1\frac{1}{2}$ by $1\frac{1}{2}$ in. The decayed wood is brown and crumbles easily. It is the cellulose that is destroyed in dry rot. White rot destroys both cellulose and lignin. It is possible for both types of rot to attack the same wood.

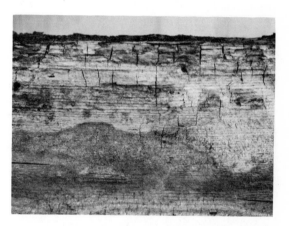

Fig. 12-9 Dry rot in Black Spruce. The uniform width of the cubes between the cracks is characteristic of this condition.

Timber preservatives are employed to protect against fungus attack and also insects. There are a great many preservatives in use. Tar oils and creosotes, familiar in railroad ties, are excellent for exposed timbers, since they have great resistance to leaching from the wood. Copper compounds, arsenic chemicals, sodium dichromate, and many other chemicals also serve as preservatives. Creosoted timbers cannot be painted. Many of the other preservatives cannot be applied to wood in contact with water because the preservative will be leached away by water.

## 12.7 lumber defects

Lumber defects are those abnormalities which influence either the strength or the appearance of the wood. Some abnormalities are not considered as defects, such as discontinuous growth rings. Some defects for one use may be advantageous for another use. Knots are defects in structural timber but may produce an attractive figure in decorative wood paneling.

*Compression wood* and *tension wood* are not uncommon. Tension wood is found on the upper side (tension side) of a leaning tree trunk and compression wood on the lower or compression side. Compression wood tends to be darker than normal wood, usually a reddish brown, and contains an unusually high lignin content. It has a higher specific gravity, greater longitudinal shrinkage, and lower toughness than normal wood, and its strength is erratic. Many failures of loaded wood members have been traced to compression wood. Like compression wood, tension wood has a higher specific gravity and greater longitudinal shrinkage than normal wood. The strength of tension wood has been found to be both higher and lower than that of normal wood. Nailing is difficult in tension wood, and sawed surfaces are fuzzy and difficult to finish.

*Checks* are longitudinal cracks that result from unequal dimension changes between the surface and the interior of the wood. The surface of the wood dries faster than the interior. Checks often follow the rays; that is, they tend to be at right angles to the growth rings. A crack that follows the growth rings is called a *shake*. Unlike checks, shakes are developed before the tree is cut.

*Pitch pockets* (resin pockets) are accumulations of pitch in a lens-shaped opening between annual rings. A *bark pocket* is a bark inclusion within the wood, A *wane* is a bark area on the edge of a board (Fig. 12.10).

A *knot* is the base of the branch within the tree. Knot-free wood is found only in the lower portion of the tree that contains no branches. If the branch

Check

Shake

Wane

Fig. 12-10 Defects in wood: check, shake, and wane.

was dead, the knot will be a *loose knot*; such a knot leaves a hole in the wood. Live branches produce an intergrown or *tight knot*, which is sound wood. If the branch is sawed along its axis, a *spike knot* is produced. The wood of a knot has a high density with much grain deviation.

*Skips* are low areas that are not smoothed by the planer knives.

The various types of warping, bow, crook, twist, and cup are illustrated in Fig. 12.11.

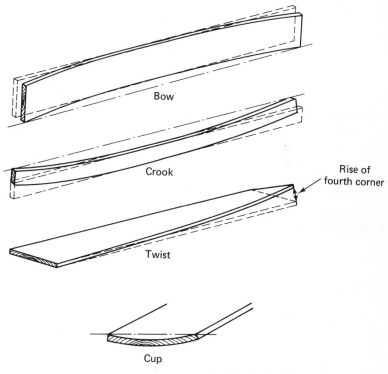

Fig. 12-11   Types of warping in wood.

## *12.8   classification*

## *and grading of lumber*

Softwood is sawed into three broad classes of lumber: yard lumber, factory and shop lumber, and structural lumber.

*Yard lumber* is that lumber used for light construction and finish work, including construction joists. It includes such items as 2 × 4s, flooring, ship-lap, siding, vee joint, moldings, and planks.

*Factory and shop lumber* is lumber in unplaned thickness, for cutting into door, sash, cabinet, and other millwork shapes. It may contain defects such as knots, since during cutting, defect areas can be discarded.

*Structural lumber* (*timber*) is the lumber intended for heavy construction, such as mine timbers, pilings, posts, joists, and planks, 2 in. thick or greater. Grading of this lumber must be based on strength, and unlike the other classes, on the use of the entire piece. All lumber must be graded for quality, but the grading systems differ in each of these three broad classes.

A large log yields boards of greatly varying quality. The varied applications in which wood is employed have different requirements, and therefore lumber is separated into various grades of quality. The grade of a piece of lumber is based on the type, location, and number of defects in the piece, such as checks, knots, and stains.

The highest grade of hardwood lumber is called Firsts, and the next grade Seconds. Both grades are usually combined into one grade, called FAS (First and Seconds). The next lower grade is Selects, then No. 1 Common, No. 2 Common, and lower grades. A brief summary of the grading the National Hardwood Lumber Association is hardly possible.

Softwood lumber is graded by a number of different associations and their rules.

Yard lumber is graded as finish lumber in grades A, B, C, D; as common boards in Nos. 1, 2, 3, 4, 5, and as common dimension in Nos. 1, 2, 3.

Finish lumber of grades A and B is suitable for natural wood finishes; grades C and D are suitable for paint finishes. Grade D may have any number of surface imperfections that do not impair the surface appearance after painting. Allowable knot diameter increases for lower grades.

Common boards contain defects that detract from appearance but are suitable for utility and construction purposes. The differences between one grade and another are due to the type rather than the number of defects. No. 1 boards may have small tight knots, such that the board is watertight. Virtually any No. 1 board will have such knots. No. 2 boards are employed for concrete forming, sheathing, and subflooring.

Dimension lumber is used for studs and rafters as well as a variety of other uses. No. 1 grade allows small knots. Defects such as stain, torn grain, or checks are not limited since they have little influence on the strength of the piece.

Structural lumber, commonly called timber, is graded for strength, since such pieces must carry loads and working stresses. An allowable working stress is assigned for each grade. The basic stress is based on the strength of clear specimens; this basic stress is reduced for each lesser grade of structural timber. The size of knots and their location, the distance that a check or a split penetrates a timber, the slope of the grain (Fig. 12.12), number of

Fig. 12-12   Slope of the grain.

Average line of the direction of fibers

annual rings per inch, and moisture content are all part of the grading. The allowable size of knot increases with the thickness of the timber. As usual for any grading of lumber, the rules are complex and may be different for different species of tree. Usually softwood structural lumber is graded as follows:

1. Dense select structural.
2. Select structural.
3. Dense construction.
4. Construction.
5. Standard.

Structural lumber is classed according to use as follows:

1. *Joists and planks:* 2–4 in. thick and at least 6 in. wide. Graded for strength in bending as a joist loaded on the edge, or as a plank loaded on the face.
2. *Beams and stringers:* 5 in. or larger in thickness, with a width of at least 2 in. greater than the thickness. Graded for strength in bending when loaded on the narrow face.
3. *Posts and timbers:* at least 5 in. or more in thickness, and a width not more than 2 in. greater than the thickness. Graded for compressive strength parallel to the grain, but not for bending strength.

The following table gives allowable stresses for Douglas fir and southern pine:

| | Bending stress Parallel to Grain (psi) | Horizontal Shear (psi) | Compression Parallel to Grain (psi) | Modulus of Elasticity |
|---|---|---|---|---|
| *douglas fir* | | | | |
| dense select | | | | |
|   structural | 2050 | 120 | 1650 | 1,600,000 |
| select structural | 1900 | 120 | 1500 | 1,600,000 |
| dense construction | 1750 | 120 | 1400 | 1,600,000 |
| construction | 1500 | 120 | 1200 | 1,600,000 |
| standard | 1200 | 95 | 1000 | 1,600,000 |
| *southern pine* | | | | |
| dense select | | | | |
|   structural | 2400 | 120 | 1750 | 1,600,000 |
| dense structural | 2000 | 120 | 1400 | 1,600,000 |
| merchantable | | | | |
|   structural longleaf | 1800 | 120 | 1300 | 1,600,000 |
| dense no. 1 structural | 1600 | 120 | 1150 | 1,600,000 |
| no. 1 dense | 1400 | 120 | 1400 | 1,600,000 |
| no. 1 | 1200 | 120 | 1200 | 1,600,000 |

## 12.9   board measure

Almost all lumber is sold by board measure, the unit being the *board foot*. Other construction materials such as board insulation may also be sold by the board foot. The board foot measures 12 in. wide, 1 ft long, and 1 in. thick. Hence there are 12 board feet in 1 ft³.

The number of board feet can be determined by end measure as follows. Consider a 2 × 4. This has 8 in.² of end measure. There are 8 board feet in 12 ft of 2 × 4. The product of the end dimensions is the number of board feet if the board is 12 ft long. The 2 × 4 may actually measure $1\frac{5}{8} \times 3\frac{5}{8}$; nevertheless it is counted as a 2 × 4, since this was its size before finishing. Lumber less than 1 in. thick is taken as a full inch for board measure.

Lumber is priced in dollars per 1000 board feet, abbreviated $/M.

## 12.10   wood fasteners

*Nails* are the usual mechanical fasteners in wood construction. Some of the many types and sizes of nails are shown in Fig. 11.28.

Nails have excellent holding power in woods, except when driven into end grain (parallel to the wood fibers) or when the nail splits the wood. The resistance of the nail to withdrawal from the wood increases with the density of the wood, the diameter of the nail, and the penetration of the nail. Of these three factors, the density of the wood has the greatest effect on holding power. Changes in moisture content of wood can greatly reduce holding power or may even cause nails to retract.

*Toenailing* is angle-driving of the nail for joints such as that of Fig. 12.13 where end-grain nailing makes an ineffective joint. Toenailing requires some skill with the hammer, since the nailhead should be buried in the wood but without severe mutilation of the wood with the hammer.

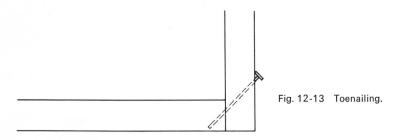

Fig. 12-13   Toenailing.

When a nail is likely to split the wood, or if the wood is too dense to be penetrated by the nail, a pilot hole smaller in diameter than the nail must be predrilled.

*Spikes* are similar to nails but are made in heavier diameters and lengths from 3 to 12 in.

*Wood screws* have round, flat, or oval heads. The flat head is used when the screw must be driven flush with the wood surface. If the wood is soft, the head will compress the wood and will be self-countersinking; dense woods require a countersink tool. Round and oval head screws are used for more ornamental applications, or when countersinking is objectionable.

In rough operations on softer woods, screws may be hammered into place rather than driven. Steel screws without protective zinc or other plating should not be used in woods subject to moisture changes, since both the fastener and the wood will rot.

*Lag screws* or lag bolts are employed for heavier wood connections. The head of a lag screw is square to receive a wrench. This fastener must be used instead of a bolt if access is limited to one side of the joint only and a nut cannot be used. The lead hole for the unthreaded shank of the lag screw must be drilled the same size as the shank. The pilot hole for the thread of the lag screw must be increased in diameter for denser woods; it may be as small as two-fifths the shank diameter for a soft wood such as white spruce, or as large as five-sixths for an oak.

For heavy timber construction, bolts, split-ring and toothed-ring connectors, and other heavy load-bearing devices are employed. The application of these fasteners is governed by established methods of timber structural design.

In the application of fasteners to wood, it must always be borne in mind that wood is weak in shear parallel to the grain (see Fig. 12.14).

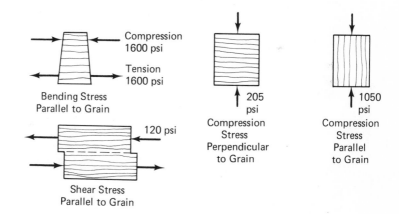

Fig. 12-14 Allowable stresses for Rocky Mountain Douglas Fir. The very low shear stress parallel to the grain is characteristic for all woods.

## 12.11 lamination of wood

Laminated timbers are familiar in construction. Many houses have a main floor beam that is laminated by spiking together several 2 × 8s or 2 × 10s to make a thick beam. Glued laminated timbers, often referred to as *glulam*, are larger structural timbers used as building frames.

A particularly important glued laminated wood product is plywood. Plywood is made of very thin laminations (veneer) with the grain of each lamination oriented at right angles to that of the next lamination. In the case of glulam, the grain direction of all the laminations is parallel to the length of the timber.

A glued wood joint should have the same strength as the wood in shear. All species of wood differ in the ease with which a satisfactory glued joint can be made. Heavy woods and heartwood are more difficult to bond satisfactorily than lightweight woods or sapwood. The best woods for bonded joints are poplar, elm, mahogany, and the softwoods western red cedar, fir, pine, spruce, and redwood.

Animal glues were originally used for glued joints in wood, but these have largely been replaced by superior types of adhesive. Animal glues give little resistance to water; casein and vegetable glues have better resistance. The best adhesives for woods (and all other materials to be bonded) are the synthetic adhesives such as the several types of formaldehyde and res-orcinol resins. Exterior grades of plywood are bonded with phenol–formal-dehyde.

Glued lumber for exterior use should have a moisture content of 10–12 per cent before gluing. Just as end grain will not hold nails satisfactorily, so the gluing of end grain is unsatisfactory as a bonded joint. A strong bonded joint requires a good fit between the two members to be bonded, since a thick glue line will give a weaker bond.

## 12.12  glulam

There is virtually no limit to the cross-sectional size or the span of a glulam member. A long-span glulam timber may have to obtain its total length from an assembly of shorter pieces of lumber, with end joints spaced apart as shown in Fig. 12.15. If the laminated timber is an arch rib, the several members are curved before laminating. For economy, low-strength woods may be used in the laminations that are not subject to high stress levels. Some

Fig. 12-15   Spacing of joints in glulam.

Stronger species

Fig. 12-16   Methods of construction in glulam.

typical glulam sections are shown in Fig. 12.16, some of these incorporating plywood in their construction.

Softwoods are used for glulam timbers. They are strong, yet lightweight, readily available and thus economical, and are generally easier to bond than hardwoods. Glue-laminated hardwoods have considerable application in small ships, sports equipment such as baseball bats, and other uses.

Casein glue and urea–formaldehyde are the most suitable adhesives for glulam. Both have some resistance to water but will not sustain prolonged immersion. Resorcinol resins are superior but more expensive. Phenol–formaldehyde is used for exterior plywood but rarely for glulam applications because of its high curing temperature.

## 12.13   plywood

Plywood is manufactured by bonding together thin sheets of wood called *veneer* in such a way that the grain direction of alternate veneers is at right angles to the previous veneer. An odd number of plies is used for balance in the mechanical properties of the plywood, for dimensional stability, and so that the grain of the faces is in the same direction. The outside veneers are called the *faces*, the central veneer is the *core*, and those plies lying between face and core are called *cross-bands*.

Most veneer is produced by the rotary cutting method shown in Fig. 12.17. The log is turned against a knife that moves toward the center of the

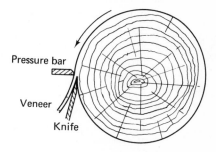

Fig. 12-17   Rotary cutting of veneer.

Pressure bar

Veneer

Knife

log at a rate which controls the thickness of the veneer. Such a peeled veneer presents a tangential surface. An alternative method is slicing (Fig. 12.18). Sliced veneers may be radial, tangential, or at any intermediate angle and can provide decorative wood figure for internal paneling.

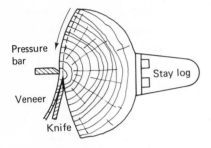

Fig. 12-18   Slicing of veneer.

In either method of cutting the sheet of wood must be bent over the knife; this bending causes small splits on the cut side parallel with the knife blade. The depth and frequency of such splits is influenced by the thickness of the veneer and the species of wood. These fine splits can appear on the face of the plywood if it is unprotected and exposed to the weather.

Several species of softwood are used in plywood, but the most common species in use is Douglas fir. Two types of softwood plywood are produced: *interior* and *exterior*. The interior type is bonded with a resin glue and can sustain occasional wet conditions. The exterior grade is suitable for permanent exterior use. Standard thickness of softwood plywood is $\frac{1}{4}$, $\frac{5}{16}$, $\frac{1}{2}$, $\frac{5}{8}$ (sheathing grade), and $\frac{3}{4}$ in. Greater thicknesses are occasionally used. Plywood sheets are $48 \times 96$ in. although longer lengths are also in use.

The following grades of Douglas fir plywood are standard:

1. *Good two sides (G2S):* both faces sound, no knots, but may contain neat patches. Waterproof glue. Suitable for highest-grade wood finishes.
2. *Good one side solid back (G/Solid):* one face sound; no knots, but may contain neat patches. The other face may contain small sound knots and has a paintable surface. Waterproof glue.
3. *Good one side (G1S):* one face sound, no knots, but may contain neat patches. The other face may have small knotholes. Waterproof glue.
4. *Solid two sides (Solid 2S):* each face solid, containing neat patches, plugs, and small sound knots. Waterproof glue.
5. *Solid one side (Solid 1S):* one face with neat patches, plugs, and sound knots. Other face may contain small knotholes and other defects. Waterproof glue.
6. *Unsanded sheathing:* both faces may have small open defects. Waterproof glue.
7. *Underlay sheathing:* similar to unsanded sheathing except that one face has a solid surface suitable for application of flexible flooring materials. Waterproof glue.

The Solid 2S and Solid 1S grades are suitable for concrete formwork. Underlay sheathing serves as a backup material for floor tile and linoleum.

Hardwood plywood is made from many species, including birch, mahogany, walnut, rosewood, and, for utility work only, poplar. The other species are principally used for their attractive appearance and figure as well as strength and hardness. Hardwood plywoods are graded into four types: Technical, type I, type II, type III. In general, the differences are based on the resistance of the adhesive bond to severe service conditions. Both Technical and type I correspond approximately to the glue bond requirements of exterior-grade Douglas fir plywood, and type II to interior grade. Type III is not required to have a water-resistant bond.

## 12.14  standard wood joints

The standard wood joints are illustrated in Fig. 12.19. Some of these joints, such as the dovetail, are restricted to cabinet work; others are also used in rough and finish carpentry.

Fig. 12-19  Standard wood joints.

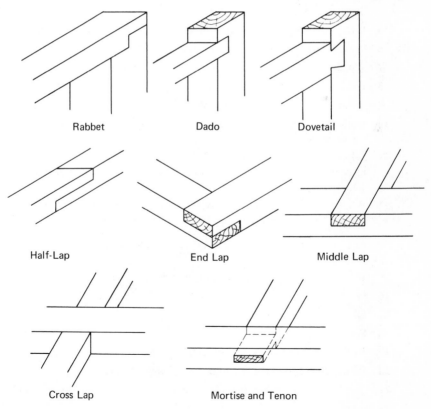

Rabbet     Dado     Dovetail

Half-Lap     End Lap     Middle Lap

Cross Lap     Mortise and Tenon

## PERTINENT ASTM SPECIFICATIONS

D2555-67 (Part 16) Establishing Clear Wood Strength Values
D143-52 (Part 16) Testing Small Clear Specimens of Timber
D198-67 (Part 16) Static Tests of Timbers
D245-64 (Part 16) Establishing Structural Grades of Lumber
D805-63 (Part 16) Testing Plywood, etc.
D1037-64 (Part 16) Properties of Wood-Base Fiberboards
D906-64 (Part 16) Strength of Adhesives in Plywood by Tension Loading
D2016-65 (Part 16) Moisture Content of Wood
D9-30 (Part 16) Terms Relating to Timber
D1038-52 (Part 16) Terms Relating to Veneer and Plywood

## QUESTIONS

1  (a) What properties of wood make it a useful building material? (b) Which of its properties are disadvantageous?

2  Name some soft hardwoods and hard softwoods.

3  Describe how you can distinguish maple from birch; walnut from mahogany; pine from spruce.

4  Sketch edge grain and cross grain.

5  What properties are required in an adhesive for glulam timber?

6  What advantage does plywood offer over sawed lumber?

7  Which is darker and harder: springwood or summerwood?

8  Distinguish heartwood from sapwood.

9  What strength properties are provided by (a) cellulose; (b) lignin?

10  With the assistance of a chemistry textbook, compare the chemical formulas for cellulose and sugar.

11  Select suitable woods for the following requirements: (a) least warping; (b) resistance to splitting; (c) ease of shaping; (d) best acceptance of decorative stains; (e) resistance to rot.

12  What is the relationship between hardness and density in woods?

13  What is the meaning of specific gravity?

14  In which direction do the greatest dimensional changes occur in wood due to moisture changes?

15  What is meant by the fiber saturation point?

16  Why does a fence post rot at the surface of the ground?

17  Distinguish between a check and a shake.

18  What is a spike knot?

19  Sketch bow, crook, twist, and cup.

20  Distinguish between yard lumber and factory and shop lumber.

21  What three factors control the holding power of a nail in wood?

It is customary to think of paper as the material for newspapers, books, business forms, and other types of printed communication. Actually paper has a remarkable range of uses in construction and other industries. As a construction material, paper is used as roofing, sheathing, vapor barrier, and various laminated products. The annual weight of paper consumed in the United States exceeds the consumption of any metal other than steel; actual paper consumption is about 50,000,000 annual tons. About 10 per cent of this total is made into various types of building papers.

The paper industry defines paper as "all kinds of matted or felted sheets of fiber . . . formed on a wire screen from water suspension." Paper is therefore one member of the important group of felt products, which include also roofing felts and certain types of wood fiberboard, such as Masonite.

Most paper is made from the cellulose fibers of wood, although other materials may sometimes be employed. Many building papers are made from waste paper and poorer grades of rags. Papermaking from wood pulp is based on the ability of cellulose fibers to adhere to each other to form continuous sheets. Paper pulp is produced by three principal processes:

# PAPER
# AND OTHER
# FELT PRODUCTS

## 13

1. The *groundwood process.*
2. The *kraft* or *sulfate process,* which produces the greatest tonnage of paper.
3. The *sulfite process.*

Groundwood or mechanical pulp is produced by grinding the wood in the presence of water. This pulp retains the brown lignin of the wood. It is used for newsprint, tissue, toweling, and board paper. Since no chemicals are required in groundwood pulpmaking, this is the cheapest paper process.

The kraft process uses alkaline caustic soda and sodium sulfide to remove the lignin from the pulp. Pulp produced by this method is strong, hence the name "kraft," which is the German word for "strong." The sulfite process produces another kind of chemical pulp by the use of bisulfites to defiber the wood.

Paper is made on a woven wire belt in a long paper machine. The wet pulp stock is pumped onto the wire belt as a thin flat ribbon from a wide nozzle. Moisture in the stock is drained through the supporting wire mesh, after which a series of pressure rolls and a dryer remove more moisture. The paper sheet is then calendered, that is, passed through a series of rolls stacked vertically. This calendering operation controls the smoothness and the thickness of the paper sheet. Finally, the sheet of paper is reeled. Many building papers, however, do not require calendering, since their smoothness is not important to their function.

Various converting operations may be performed on the paper stock to produce the final product. The word "converting" refers to operations that convert semifinished products such as paper into manufactured articles. The converting process may saturate the paper with asphalt, or coat it with asphalt or wax. Lamination provides a paper with a combination of properties or surfaces. The lamination of paper with aluminum foil converts the paper stock into a suitable vapor barrier, the paper serving to protect and strengthen the thin aluminum foil. In many types of laminated building papers, asphalt is used as the bonding medium to join the laminations, which often are two sheets of kraft paper. About 1 lb of asphalt is used per 100 ft$^2$ of paper.

Corrugated paperboard is made of three webs of kraft paperboard. The middle one is corrugated by means of corrugating rolls, the outer laminations being coated to it with a suitable adhesive, such as starch. The corrugated sandwich is highly resistant to bending, crushing, or even penetration. Larger structural paper sandwiches are made by using a thicker honeycomb paper core bonded to suitable facing materials.

## 13.2 construction papers

Two types of *sheathing paper* are available. Plain sheathing papers are usually tough kraft papers produced in several weights from about 4 to 10 pounds per square (100 ft²). Asphalt-impregnated and coated papers are felt and kraft papers in various weights from 4 to 10 pounds per square. These are vapor barriers. The usual width for sheathing papers is 36 in.

*Cushioning paper* is a wood-fiber insulating paper used as a cushion under carpets and linoleum. Fireproofing paper, used to reduce the spread of fire, is made from asbestos fibers. As mentioned in Chapter 9, gypsum wallboard is coated with paper, actually kraft paper.

*Vapor-barrier papers* are commonly used in construction. These are available in rolls 36 in. wide, in a variety of types. One is a waxed kraft. A second type consists of two laminations of paper with a bonding and barrier film of asphalt. A third type is a sheet of kraft laminated to copper foil with asphalt, used both for vapor barrier and for flashing purposes.

Cylindrical concrete forms (Fig. 13.1) are made of heavy paperboard wound in a spiral tube and wax-coated. These are available in various diameters from 6 to 48 in., with a length of 12 ft. They are useful for pouring concrete columns or for forming ducts in concrete floors.

Fig. 13-1 Spiral paperboard tube for concrete forming.

## 13.3 roofing papers

Roofing papers fall under two broad classes: *roofing felts* for built-up roofs, and *rolled roofing*, a heavy paper with a mineral surface used as a final roofing membrane. Both are produced from a groundwood pulp of long fibers and sufficient porosity for saturation with asphalt. Rolled roofing is either 18 or 36 in. wide in various weights. Roofing felts are 36 in. wide in rolls, also in various weights. Weights of roofing materials are expressed per square, which is 100 ft².

Roofing felts are saturated by repeated dips into a vat of molten asphalt. A uniform distribution of asphalt is essential. The finished saturated felt must not be brittle or so tacky that two sheets will adhere and be torn if pulled apart.

Some roofing felts are made of asbestos or fiberglass fibers; asbestos fibers are more expensive, however. The fiberglass felt offers the advantage of a greater absorption for asphalt.

In a roofing felt the asphalt is the waterproofing agent. The felt reinforces the asphalt and prevents it from sagging under the influence of high roof temperatures. It also prevents cracking of the asphalt in cold weather.

## 13.4 building fiberboards

*Building fiberboards* are sheets made of fiber, usually wood or asbestos. Binding agents may be added, such as asphalt. Such boards are used for insulation, panels, and other applications. The lightweight types are insulating and roofing boards, and the higher densities serve as structural and finishing surfaces. A variant of the fiberboard is the resin-bonded particle board, which is made by binding wood particles into a flat board with a resinuous binder such as urea-formaldehyde. The edges of these boards may be plain, ship-lapped, tongue-and-groove, vee-joint, or other. The principal types are general-use boards, insulating board, roofing boards, interior board, sheathing board, and hardboard.

In addition to these types, there are asbestos fiberboards made of asbestos fiber and cement. The asbestos board is used for fireproofing and acoustical absorption.

The usual size of all these types of fiberboard is 4 × 8 ft.

## 13.5 hardboard

*Hardboard* is made from wood fiber pressed into a hard sheet on a heated press to a density of about 60 pcf. The front face is smooth and brown in color; the back face shows the impression of the screen on which it was felted. The screen is necessary to allow steam to escape from the board when it is hot-pressed. It is possible, however, to produce such a hardboard with two smooth surfaces, by a modification of the manufacturing process.

Hardboards are classified as either standard (untempered) or tempered. The *untempered* type is not suitable for exterior use. The *tempering* process, which produces a dark brown color, consists of impregnation of the pressed board with drying oils, followed by baking. This treatment increases the strength and the moisture resistance of the board and results also in a heavier board.

Hardboard, of which Masonite is the most familiar of the brand names, is manufactured in $\frac{1}{8}$- and $\frac{3}{16}$-in. thicknesses. The usual length is 8 ft. The boards are smooth face, can be bent to a curvature, and are adaptable to many kinds of wood finishes. In many applications they compete with plywood. They are sometimes employed for reusable concrete forms, when a smooth concrete surface is required. Some hardboards are prefinished wall panels, with a factory-applied finish and vertical grooving to simulate wood joints. Such hardboard panels are especially successful for rooms such as basement rooms, where there are conditions of high humidity that might buckle other types of panels.

## 13.6 other types of building board

Other types of wood board include the resin-bonded wood particle board and chipboard. *Particle board* has similar characteristics and applications to wood-fiber hardboard, although it may shrink and swell more with humidity changes. *Chipboard* is made of very thin wood chips about $1\frac{1}{2}$ in. square. Particle board uses fine wood particles larger than sawdust, which are coated with an adhesive, pressed and cured. Particle board is not used for exterior work or interior finish, since it does not weather well or have an attractive finish. Principal uses are floor underlay, partitions, shelves, and cores for panel finishes.

Brief mention will now be made of some of the many other types of building board—made of paper, straw, mineral fiber, and cork. Gypsum board was described in Chapter 9.

*Strawboard* is made of compressed straw with a kraft-paper surface, in both a structural and an insulating board. The structural board is a type of partition panel board for non-load-bearing partitions, suitable as a plaster base where necessary, but it has also been used for roof insulation. Insulating strawboard is used only as a roof insulation.

*Corkboard*, made of cork, is used as an insulating board, with occasional applications as an interesting interior wall finish.

*Paperboard* is made either of pressed paper pulp $\frac{3}{16}$ or $\frac{1}{4}$ in. thick, or of corrugated paperboard.

*Mineral fiberboard* consists of mats of rockwool or fiberglass with paper faces in a stiff construction, used only for roof insulation. Such boards are bonded to the roof with asphalt.

The polystyrene and polyurethane rigid-foam insulating boards are discussed in Chapter 15.

## 13.7 asbestos-cement board

Asbestos is a mineral with a unique property: its fibrous structure. This and its fire resistance make it a valuable material for construction. *Asbestos-cement board* is a rigid board made of portland cement with 10–15 per cent asbestos fiber added and steam-cured. Both flat and corrugated board is manufactured, in thicknesses of $\frac{1}{8}$, $\frac{3}{16}$, $\frac{1}{4}$, $\frac{3}{8}$, and $\frac{1}{2}$ in. The white corrugated sheets are frequently seen on buildings as wall cladding, roof decks, and even roofing (Fig. 13.2).

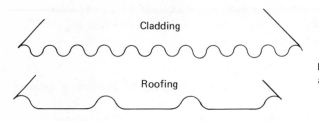

Cladding

Roofing

Fig. 13-2 Types of corrugated asbestos-cement board.

Insulating asbestos board uses a core of wood fiber, fiberglass, or foamed polystyrene in a sandwich construction with faces of asbestos–cement.

Asbestos–cement is hard and abrasive to tools; it requires metalworking tools, such as hacksaws. Asbestos board should be drilled and not punched. Special types of fasteners are used to secure the corrugated asbestos board.

## QUESTIONS

1 State the technical definition of paper as used by the paper industry.

2 Paper is a felt product. Explain why.

3 What is kraft paper?

4 Why does newsprint turn brown or yellow with age?

5 What is meant by calendering?

6 In an paper–asphalt combination, what properties or functions are contributed by each member of the pair?

7 Define a square.

8 What is a roofing felt?

9 In a roofing felt, what functions are supplied by the felt and by the asphalt?

10 What is the difference between tempered and untempered hardboard?

At the time of writing, over 2,000,000 tons of plastics and rubbers are consumed annually by the construction industry in the United States. This consumption is expected to double by 1980.

The polymers are long-chain organic chemical materials based on carbon, those of greatest interest to the construction industry being wood, paper, asphalts, the plastics, and the rubbers. The woods have enjoyed a long and honored history as construction materials, and although now meeting strong competition from plastics, wood, with its unique properties and infinite variety of finishes and grain patterns, would be abandoned with great reluctance. Paper has been used in construction on this continent for a hundred years, and longer in Asia.

The plastics were introduced into construction in earlier decades of this century with great optimism and frequent disaster. The early plastic materials were of doubtful quality and presented problems of weathering, stability, and durability that were not well understood. The employment of plastics in the construction industry is now characterized by much more conservatism, experience, and knowledge, and the quality of modern plastics materials is often superb. Nevertheless, their successful employment requires a basic knowledge of their special properties. For example, polyvinyl chloride,

# POLYMER
# MATERIALS
# IN CONSTRUCTION

# 14

generally called simply *vinyl*, has excellent weathering qualities if suitably compounded for such characteristics, but not every grade of polyvinyl chloride is suitable as an exterior building material. There may be as many as a hundred different formulations of some plastic materials, polyurethane for example, and there are already at least a hundred basic polymeric materials, with new types constantly under development. Just as there is no such material as "steel," there are no such materials as "plastic" and "rubber." Faced with so many possibilities, the designer and specifier must be acquainted with the basic fundamentals and concepts that can make an organization out of the chaos of information and facts. Fortunately, the technology of the polymers is both easy to understand and interesting.

Fig. 14-1   The largest polyethylene extruder ever built in Canada. This extruder produces industrial film at Union Carbide Canada Limited's Lindsay, Ontario plant. The 8-in. extruder can make film 40 ft wide and has a capacity in excess of 1500 lb/hr. (Photo courtesy Union Carbide Canada Limited, Toronto.)

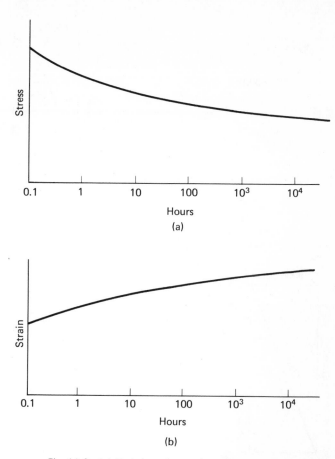

Fig. 14-2 (a) Variation of stress in a plastic with time; strain held constant (stress-relaxation curve); (b) variation of strain in a plastic with time; stress maintained constant (creep curve).

## 14.1 four basic types of polymers

In earlier years the polymers were separated into two broad classes, plastics and rubbers (elastomers). Rubbers were elastic, or bounced; plastics deformed plastically. There were many inconvenient exceptions to this simple classification: Bakelite (phenol–formaldehyde) does not strain plastically, even though it is called a plastic. This classification is no longer helpful. Polyurethane can be formulated for either rubber or plastic properties. Polycarbonate, a glazing substitute for glass in windows, also behaves both as a rubber and as a plastic. Epoxies, formaldehydes, and many other plastics are incapable of plastic deformation, nor do they have the extensibility of elastomers.

Before sorting these many materials into their basic types, the terms "thermoplastic" and "thermosetting" must be understood. A *thermoplastic*

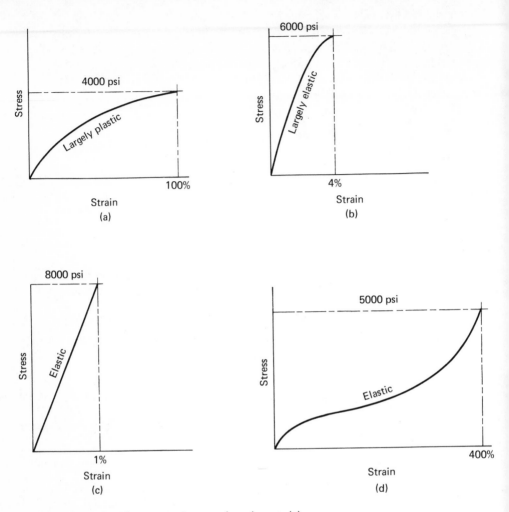

Fig. 14-3 Typical tensile stress-strain curves for polymers: (a) flexible thermoplastic; (b) rigid thermoplastic; (c) thermosetting plastic; (d) elastomer. These do not disclose any tendency for creep under sustained loads, nor can behavior in compression be predicted from these curves. Values given for maximum stress and strain are typical ones.

Fig. 14-4 Flame spread sample.

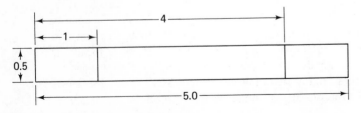

material can be repeatedly softened or made to flow by heating. A *thermosetting* material will not soften when heated; it may char, or ignite, but it will not soften. Familiar thermoplastics are vinyl, polyethylene, and asphalt; familiar thermosets are wood, paper, rubber, and Bakelite. With the use of these definitions we can now distinguish four types of polymeric materials:

1. *Flexible thermoplastics*, such as polyethylene, capable of large plastic deformation.
2. *Rigid thermoplastics*, such as polyvinyl chloride and polystyrene, limited in their maximum possible strain and therefore brittle.
3. *Rigid thermosets*, such as epoxy and phenol–formaldehyde (Bakelite), also brittle. All thermosetting plastics are brittle.
4. *Elastomers or rubbers*, distinguished by remarkable elastic extensibility.

All four types of polymers exhibit considerable creep and stress relaxation under conditions of prolonged stress (Fig. 14.2). In a standard stress–strain test, each of the four types gives a characteristics shape of stress–strain curve (Fig. 14.3). There are also exceptional types of behavior. The stress–strain curve for polyurethane (Fig. 1.6) does not correspond closely with any of the characteristic curves of Fig. 14.3.

## 14.2 becoming acquainted
## with the polymers

There is no quicker method of becoming familiar with the polymer materials and their characteristics than to make a flammability test on them. A flammability test can be set up to follow more or less closely the procedure of ASTM D635-63.

Test samples of plastic sheet are cut 5 in. long and 0.5 in. wide, and a sharp pencil mark scribed across each sample at 1 in. and 4 in. from one end of the sample (Fig. 14.4). At least five samples of each polymer should be tested. The samples are mounted at a 45-degree angle with axis horizontal (see Fig. 14.5). The free end of the sample is ignited with a bunsen burner or other suitable ignition means. The time for the specimen to burn from one pencil mark to the other is found from a stopwatch or a sweep-second hand on a wristwatch, and the average time for all specimens of one type is determined. If the sample does not ignite after two attempts at ignition, it is designated "self-extinguishing." It is also designated "self-extinguishing" if it ignites but does not continue to burn after the igniting flame is removed.

All pertinent data should be recorded, including thickness, and for foamed plastics the density in pounds per cubic foot must be known. In

Fig. 14-5  Flame spread test on a sample of polystyrene.

recording results, note also the character of the flame, since a flame test is the most useful method of identifying unknown plastics.

As wide a variety of polymers as possible should be tested, including those not currently used as construction materials. In the case of foamed polystyrene (styrofoam), obtain both a standard and a flame-retardant type (a sample from a styrofoam coffee cup will do for the untreated type). If only one type is tested, it is possible to be led to conclude that all polystyrene is flame-retardant (or the opposite). Similarly, as many formulations of polyurethane foam as possible must be tested to avoid drawing erroneous general conclusions. Some polyurethanes are self-extinguishing and some burn readily.

The usefulness of such a test is greatly extended if it is applied to thin samples of the many woods, papers, and board stock used in construction, and if results on these materials are compared with the results with plastics and rubbers. If possible, obtain a sample of cellulose nitrate for this flame test to see why the material is totally unacceptable as a construction material.

The established construction polymers are the following:

1. Various woods.
2. Kraft paper.

3. SBR, butyl, and polychloroprene (neoprene) rubbers.
4. Asphalts.
5. Chlorosulfonated polyethylene (Hypalon).
6. Unplasticized (that is, rigid) polyvinyl chloride (PVC).
7. Foamed and solid polystyrene (PS).
8. Foamed polyurethane, densities 2 and 4 pcf.
9. Acrylonitrile–butadiene–styrene (ABS).
10. Polymethyl methacrylate (PMMA, Lucite, Plexiglas).
11. Polycarbonate.
12. Foamed and sheet polyethylene (PE).
13. Polyvinyl fluoride (PVF, Tedlar).
14. Epoxy.
15. Foamed urea–formaldehyde.
16. Phenol–formaldehyde (Bakelite).
17. Polyesters (in reinforced plastics and Mylar drafting paper).

### 14.3 chemical structure

### of thermoplastics

Most of the plastics, rubbers, and asphalts used in the construction industry are manufactured from petroleum hydrocarbons extracted from petroleum and natural gas. A hydrocarbon is a chemical compound of carbon and hydrogen, and most of the petroleum and natural gas materials, such as gasolines, are hydrocarbons. Many polymers are derived from wood, including cellulose acetate, cellulose nitrate, cellulose acetate butyrate (CAB is used for handles of hacksaws and screwdrivers), and others, but these cellulose polymers are of minor interest in construction, with the exception of wood itself and its paper derivatives.

The number of hydrocarbon chemicals in petroleum is probably infinite. The lighter components, such as methane, ethane, and propane, can be separated out individually, but the great multitude of compounds are in the heavier liquid and semisolid fractions, which are made into asphalts. To illustrate the number of possible hydrocarbons, consider the common gasoline hydrocarbon, octane. This is the petroleum component that gave its name to the octane rating of gasoline. It contains 8 carbon atoms and 18 hydrogen atoms. These 26 atoms can be rearranged chemically to form 17 other compounds. By reducing the number of hydrogen atoms attached to the 8 carbon atoms, still more hydrocarbons are possible. A selection of octane compounds is given in Fig. 14.6. Twenty-five carbon atoms offer the possibility of about 40,000,000 different hydrocarbons, and 25 is by no means the maximum number of carbon atoms in any petroleum molecule.

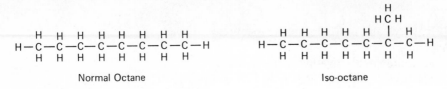

Normal Octane           Iso-octane

Fig. 14-6  Two types of octane. By various branching arrangements many other types are possible.

Fig. 14-7  Chemical compounds in natural gas and petroleum.

| Formula | | Name of chemical | Condition | Use |
|---|---|---|---|---|
| 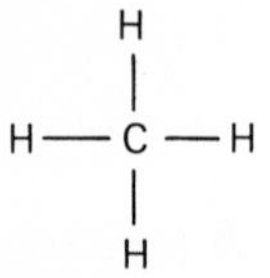 H—C—H with H above and H below | $CH_4$ | Methane | Gas | Heating fuel |
| 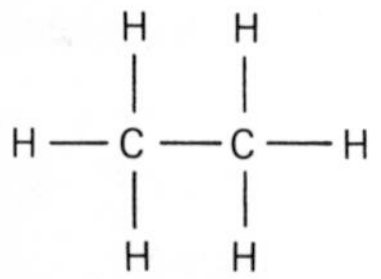 H—C—C—H with H's | $C_2H_6$ | Ethane | Gas | Converted into plastics |
| 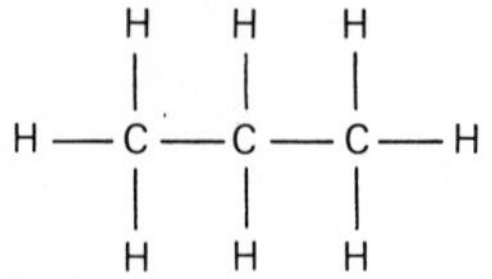 H—C—C—C—H with H's | $C_3H_8$ | Propane | Gas | Heating fuel |
| 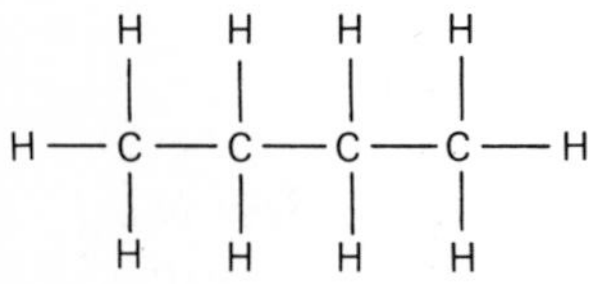 H—C—C—C—C—H with H's | $C_4H_{10}$ | Butane | Gas | Heating fuel or converted into rubbers |
| $C_5H_{12}$ $C_6H_{14}$ $C_7H_{16}$ $C_8H_{18}$ $C_9$, etc. | | Pentane Hexane Heptane Octane | Liquid | Converted into engine fuels (gasoline) |
| $C_{100}$, etc. | | Asphalts and tars | Solid | Road and roofing asphalts |

It is this remarkable flexibility of carbon chemistry that explains why there are so many plastics and rubbers and why so many more will be invented. Any polymer chemist can invent another new plastic formula every morning for a year, although all these prospective formulas would not necessarily be suitable plastics or rubbers for the construction industry.

Most of the hydrocarbons in petroleum can be classified into four groups: paraffins, olefins, naphthenes, and aromatics. The naphthenes will not be discussed here.

1. *Paraffins:* have the general formula $C_nH_{2n+2}$. The simpler paraffins are shown in Fig. 14.7. These are straight-chain hydrocarbons. The names of the paraffins end in *-ane*: methane, ethane, propane, and so on. These are saturated hydrocarbons, the term "saturated" meaning that they contain the maximum possible number of hydrogen atoms in each case. For this reason the paraffins are relatively inert chemically.

From Fig. 14.7 it will be noted that carbon has four linkages to other atoms, or in chemist's language, a valence of four. Hydrogen has a valence of one, oxygen two. In inventing new plastics formulas, these bonding relationships must be followed.

Paraffins with 4 or fewer carbon atoms are gases at ordinary temperatures, those with 5–15 carbon atoms are liquids, those with over 15 are waxes, asphalts, and so on. This general principle is true for plastics; those with only a relatively few carbon atoms in the chain are liquids, while the solid plastics and rubbers are made of very long chains of carbon atoms.

2. *Olefins:* these (Fig. 14.8) are unsaturated. If two hydrogen atoms can be added to an olefin to saturate it, it is a mono-olefin, $C_nH_{2n}$. The mono-olefins have names that correspond to those of the paraffins to which they are related, with the ending *-ene* or *-ylene*, such as propylene and ethylene. If four hydrogen atoms can be added to produce saturation, the olefin is a diolefin, $C_nH_{2n-2}$, with a name ending in *-diene* (pronounced dye-een), as butadiene. Olefins are chemically reactive and serve as the principal raw materials for the manufacture of plastics, rubbers, and a range of other materials, such as ethylene glycol. Note that they have the straight-chain structure of paraffins, but because of unsaturation, some of the carbon atoms are held to each other with double bonds.

Fig. 14-8 Olefins.

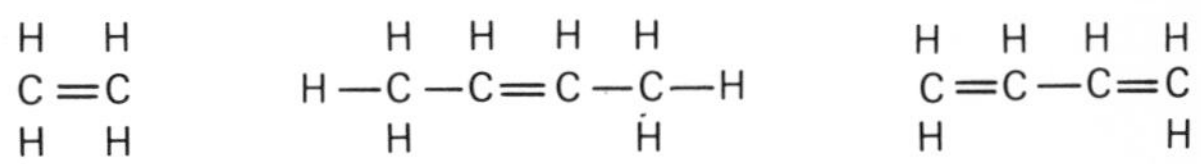

| Ethylene | Butylene | Butadiene |

3. *Aromatics:* ring-shaped compounds like the naphthenes but unsaturated. Thus they contain some double bonds between carbon atoms. Figure 14.9 shows the structure of benzene, a typical aromatic. Since they are unsaturated, the aromatics are chemically active and are the materials for the manufacture of a wide range of organic chemicals, including explosives, solvents, and polystyrene.

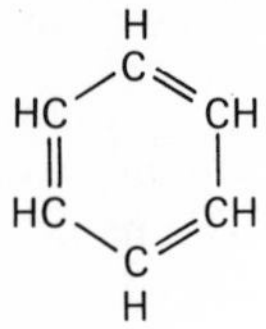

Fig. 14-9  Benzine, an aromatic.

Only limited amounts of olefins are found in crude petroleum. The heavier paraffins can be used in asphalts, and the lighter paraffins are used as heating and engine fuels after being refined. Ethane, however, is usually

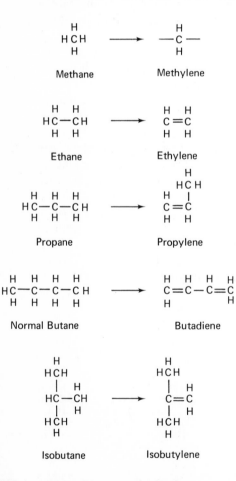

Fig. 14-10  Conversion of light paraffins to unsaturated olefin intermediates.

reserved for conversion into *ethylene*, a chemical intermediate from which such plastics as polyethylene can be made. Butane is converted into butylene or butadiene, the intermediates from which rubbers such as butyl rubber are made.

We can now consider the method of formation of a few of the simpler plastics, those which use the lightest of the petroleum and natural gas compounds, methane, ethane, propane, and butane. The formulas of these four compounds are given in Fig. 14.10.

To make each of these into its own polymer, it must be first converted into an olefin. Note that every carbon atom must maintain four linkages to other atoms.

To make the four liquid intermediates into solid plastics and rubbers, they must be polymerized, joined into long continuous chains. If about 500–1000 ethylene monomers are joined in sequence, we have the familiar polyethylene (Fig. 14.11).

Of the polymers shown in Fig. 14.10, polyethylene has useful properties and is inexpensive. Polypropylene is another useful plastic, but since it does not weather well, it is not used in construction. Polymethylene was tested about 75 years ago and was found to be brittle, despite the close resemblance of its formula to that of the flexible polyethylene. Polybutadiene and poly-

Fig. 14-11   (a) Ethylene and polyethylene; (b) variation of tensile strength with length of carbon chain in a typical polyethylene.

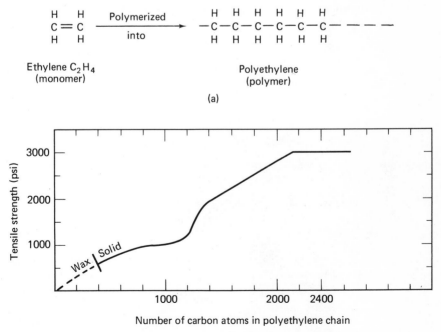

isobutylene are two of the elastomers used as sealants for glazing and curtain wall construction.

Although the polymeric materials are usually diagrammed as straight chains, actually there is some angularity between adjacent carbon links. The high elongation of many of these materials is explained by the straightening out of these angular links. Also, there are always some short side chains attached to the main carbon chain.

## 14.4 common thermoplastics

Polyethylene is a thermoplastic material. From its basic formula it is possible to invent, on paper at least, a very large number of new thermoplastics, being careful to observe the correct number of bonds associated with any element.

Suppose that we wish to invent a plastic rain gutter for houses (this of course has already been done). Polyethylene is an unsuitable material for a number of reasons, one being its flammability. Any chemical material made up of strings of carbon and hydrogen atoms is certain to be an excellent fuel. If no building code should allow polyethylene for this purpose, then the ethylene monomer can be modified to be nonburning. The element chlorine does not burn. By replacing at least one hydrogen atom by chlorine in the monomer, a nonburning thermoplastic material results. Chlorine bonds once, like hydrogen. The modified monomer, polyvinyl chloride, is called *vinyl*; see Fig. 14.12.

$$
\left[\begin{array}{cc} H & H \\ C - C \\ H & Cl \end{array}\right]
$$

(a)

(b)

Fig. 14-12   (a) The monomer (vinyl chloride) of polyvinyl chloride; (b) an injection-molded rainwater fitting, gutter to downspout, made of rigid polyvinyl chloride.

Polyvinyl chloride proves to be an excellent material for a rain gutter. It does not burn, and unlike polyethylene, it is a stiff plastic. It is also one of the lowest priced.

If we substitute two chlorine atoms, we obtain polyvinylidene chloride also called by its trade name, Saran (Fig. 14.13). This thermoplastic is not used in construction, however. The addition of chlorine atoms can be carried to the maximum of four.

Fig. 14-13   Polyvinylidene chloride.

$$-\overset{\text{H}}{\underset{\text{H}}{\text{C}}}-\overset{\text{Cl}}{\underset{\text{Cl}}{\text{C}}}-$$

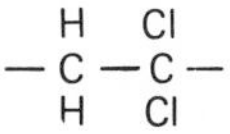

The element fluorine, which has a single bond, like hydrogen, also does not burn. One hydrogen atom in the ethylene monomer can be replaced by fluorine to yield the polyvinyl fluoride of Fig. 14.14. PVF is a thermoplastic of outstanding properties, nonburning, and used as a surface film to protect other construction materials from weathering, ultraviolet degredation, corrosion, or damage by scuffing. It is sold under the trade name Tedlar.

Fig. 14-14   Polyvinyl fluoride.

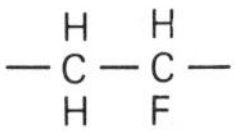

$$-\overset{\text{H}}{\underset{\text{H}}{\text{C}}}-\overset{\text{H}}{\underset{\text{F}}{\text{C}}}-$$

Two fluorine atoms give polyvinylidene fluoride, a construction film with properties similar to Tedlar, although it is less well known. Four fluorine atoms produces the famous Teflon, polytetrafluoroethylene. A great many other monomer possibilities can be designed using fluorine and chlorine or both, but none is in use as a construction material.

Clearly the chemist, or even the reader of this book, can continue to "invent" thermoplastics almost without end. Each formula so easily invented would of course have to be manufactured and tested for properties, and many would be found unsuitable. The other linear thermoplastics that have established a place for themselves in construction materials are given in Fig. 14.15.

A more complex linear thermoplastic is ABS, acrylonitrile–butadiene–styrene, a copolymerization of three monomers. ABS is a modification of polystyrene. A pure polystyrene plastic would be unsuited to construction applications by reason of its brittleness and its poor resistance to ultraviolet degredation. Modified types of polystyrene must be used in construction products, even for indoor use. Another solution is to include some butadiene in the formulation. Butadiene is the rubber polymer from which automobile tires are made, and it provides considerable toughness to the ABS polymer. If properly compounded and wisely used, ABS is suitable even for exterior use in buildings.

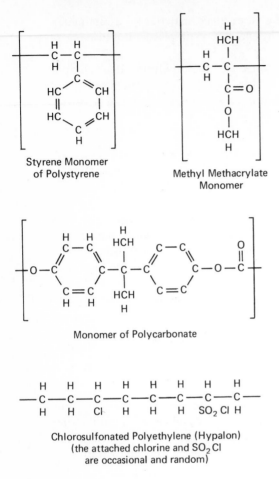

Styrene Monomer
of Polystyrene

Methyl Methacrylate
Monomer

Monomer of Polycarbonate

Chlorosulfonated Polyethylene (Hypalon)
(the attached chlorine and $SO_2$ Cl
are occasional and random)

Fig. 14-15   Construction thermo-plastics.

## 14.5   applications of

## linear thermoplastics

ABS, polystyrene, polyethylene, and PVC are used in the construction indus-try both as solid materials and as foams. The discussion of plastic foams is deferred to Chapter 15. Except for polyethylene, which is too soft and waxy, all these materials can be made to simulate wood in moldings or paneling of complex shape. They can imitate wood grain, dark wood colors, and even the open grain of hardwoods. Such wood substitutes can often be found in the decorative interior finishes of restaurants along the interstate highway system.

**PVC.** *polyvinyl chloride* is a hard and slightly brittle material when un-plasticized. When suitable ultraviolet stabilizers are incorporated, it has excellent weathering and ultraviolet resistance. Softer grades of PVC incorporate plasticizers, as for example in such products as garden hose. The plasticized and soft grades have poor weatherability and previous attempts to use them in exterior applications have usually been unsuccessful. The plasticizers usually leach out of the material under extended outdoor exposure. PVC is brittle in severe cold; ABS is slightly less brittle at low temperatures.

Unplasticized PVC, abbreviated UPVC, is used as a colored surface finish on exterior steel panels of prefabricated steel buildings and on aluminum house siding, on curtain wall surfaces, and in various extruded shapes such as imitation wood moldings and vinyl house siding.

Polystyrene and polyethylene are unsuited to exterior applications. Ultraviolet radiation from the sun will depolymerize these materials and also cause them to oxidize. Polyethylene pipe is blended with carbon black, the purpose of the black being the absorption of ultraviolet radiation for protection of the polymer.

**Polyethylene.** The construction industry uses *polyethylene* as film for concrete curing and protection, for cold-weather hoardings, for protection of stored building materials from rain and snow, as a vapor barrier, and a number of other similar uses. Polyethylene film is available in three colors, clear, white, and black; and in 2-, 4-, and 6-mil and other thicknesses (1 mil = 0.001 in.).

**ABS.** *Acrylonitrile–butadiene–styrene* is a copolymer of three monomers for toughness. Its chief uses in construction are for molded panels, imitation wood moldings, and sheet material. This material and polystyrene are also used for many fluorescent lighting fixtures. ABS is not readily distinguishable from the high-impact grades of polystyrene.

**PMMA.** Although there are a number of acrylic plastics, the construction industry uses *polymethyl methacrylate*, often referred to by brand names such as Lucite and Plexiglas. This material is employed as a substitute for glass in lighting fixtures, skylights, and windows. These are crystal-clear plastics of high rigidity and hardness, with slightly better light transmission than glass, about 90 per cent. They are somewhat expensive, but replace glass in school buildings where replacement of broken windows becomes expensive. Acrylics are easily scratched, however.

**Polycarbonate.** *Polycarbonate* is a remarkable plastic, hard, strong, and incredibly tough. It has been used for nuts, bolts, screws, nails, and hard hats. Like the acrylics, it replaces glass where vandalism is a serious problem. Glass is readily smashed, acrylic may be fractured with some effort, but a sheet of polycarbonate glazing can be removed only by sufficient brutality to knock the sheet completely out of its frame. Firemen do not like this

## CONSTRUCTION THERMOPLASTICS

| | ABS | Polyethylene | UPVC | Polystyrene | Acrylic | Polycarbonate | PVF |
|---|---|---|---|---|---|---|---|
| specific gravity | 1.04 | 0.95 | 1.4 | 1.05 | 1.2 | 1.2 | 1.5 |
| tensile strength (psi) | 5000 | $2\text{--}4 \times 10^3$ | 6000 | 5000 | 8000 | 9500 | 18,000 |
| elongation (%) | 40 | 40–400 | 10 | 1.2 | 5 | 100 | 200 |
| modulus of elasticity | $250 \times 10^3$ | $25\text{--}100 \times 10^3$ | $350 \times 10^3$ | 50,000 | $400 \times 10^3$ | $350 \times 10^3$ | $300 \times 10^3$ |
| impact strength | good | good | poor | poor | low | high | high |
| thermal expansion (per °F) | 0.00005 | 0.0001 | 0.00003 | 0.00004 | 0.00005 | 0.00004 | 0.00003 |
| resistance to heat (°F) | 160 | 200 | 150 | 160 | 150 | 250 | 120 |
| burning rate, in./min. | $1\frac{1}{2}$ | 3 | none | 10 | 2 | 1 | slow |
| effect of sunlight | slight | serious | slight | serious | slight | slight | none |
| type of thermoplastic | flexible | flexible | rigid | rigid | rigid | flexible | flexible |

material because it prevents entrance into a burning building through the windows. Because of a higher expansion coefficient than glass, some redesign of the window frame and sealing is usually necessary when glazing with acrylic or polycarbonate.

**PVF.** *Polyvinyl fluoride* more commonly known by its trade name Tedlar, is perhaps the most inconspicuous of all construction materials. It is supplied only as film in thicknesses of $\frac{1}{2}$–3 mils. Its use is largely confined to that of a scuff-proof, soil-proof, and weatherproof surface film on interior surfaces such as decorative wood panels, or on roofs or exterior walls. For such protective applications its performance is superb, much superior to paint. It acts as a filter against ultraviolet radiation. For roofing application it is not applied over built-up roofs, but over the newer types of roofing such as Hypalon or sprayed polyurethane.

Figure 14.15 showed a polyethylene modified with sulfur and chlorine. This is chlorinated polyethylene, also known as Hypalon. Hypalon will be discussed later under the topics of rubber and roofing.

## 14.6 properties of construction thermoplastics

The accompanying table is a summary of the properties of the thermoplastic materials used in the construction industry. Since all these basic thermoplastics are available in several formulations, the data supplied in the table are only approximate. Burning rate and impact resistance especially will vary with the formulation. Again, a thermoplastic supplied as film is stronger than the same plastic in a solid section. The table nevertheless presents a useful summary and comparison of the types of thermoplastic materials.

The values given for ultimate tensile stress are those obtained in the usual short-time tension test. Such a test indicates the relative ability of the thermoplastic to sustain momentary loads only, but allowable stresses for sustained loads must be taken well below the ultimate tensile strength to allow for creep at high stresses. The ultimate strength of almost all thermoplastic materials will also decrease slightly over the years. This decrease in strength would not exceed 20 per cent for a thermoplastic, unless it is a type unsuited to outdoor exposure and exposed to exterior conditions.

The thermal expansion of the thermoplastics is about 10 times that of steel, and $E$ values about $\frac{1}{100}$ that of steel.

The effect of sunlight (ultraviolet radiation) on thermoplastics is difficult to summarize. The table says that sunlight has little effect on unplasticized polyvinyl chloride, a statement that is not true for pure polyvinyl chloride.

But commercial grades of unplasticized polyvinyl chloride contain pigments and ultraviolet absorbers such as titanium dioxide, and in such formulations the material is one of the most weather-resistant of the thermoplastics. Pigments and coloring matter have a very favorable influence on the ultraviolet stability of these materials. Carbon black is by far the best additive for ultraviolet protection, but it cannot be used if the plastic must be colored.

Degredation of thermoplastics due to weathering and sunlight will result in crazing, cracking, yellowing, brittleness, and hardness. Ultraviolet radiation and oxidation also will in time degrade the asphalts.

## 14.7 asphalts

The *asphalts* are thermoplastics composed of long chains of carbon atoms. Although they are modified by oil-refining processes, nevertheless the asphalts are natural polymers like wood. Wood is thermosetting, whereas asphalts are thermoplastic.

Asphalts are the residua or leftovers after the lighter constituents of petroleum are distilled away. There are also some natural rock asphalts, which are calcareous types of rock impregnated with bitumen. The words "bitumen" or "bituminous" embrace both the petroleum and the rock material, both of which are soluble in carbon disulfide. Coal tars, also called pitches, serve much the same general purposes as the asphaltic materials, but are extracted from coal. All these materials, whatever their origin, are chiefly used for waterproofing: as damp-proof courses and membranes both below grade and on roofs, and for impregnating building paper, roofing felts, and building board. They are used also in floor tile, asphalt shingles, and bituminous and aluminum paints. When incorporated with fine aggregate, they are used for the surfacing of roads and paved areas. The adhesiveness of these materials to most surfaces contributes greatly to their usefulness. ASTM D8-55 defines the bituminous materials in more exact terms.

These bituminous materials are mixtures of thousands of different petroleum or coal compounds, including paraffins, olefins, and aromatics, in straight and branched carbon chains. They include in their chemical structure oxygen, nitrogen, sulfur, nickel, vanadium, and traces of other elements.

The viscosity and flow properties of these bituminous thermoplastic materials are significant for all their applications. A roofing asphalt applied on a steep roof must not sag under conditions of high summer temperature and direct radiation from the sun. The usual test for viscosity or consistency is the penetration test. A weight of 100 g is applied to a penetrating needle, and the depth of penetration of the needle into the asphalt is measured after 5 sec using a temperature of 77°F (25°C). The test is similar to a hardness test for metals, rubbers, or plastics, since it is based on a penetration principle.

An alternative test is a test for softening point, using the *ring-and-ball method*. The temperature is measured at which a steel ball $\frac{3}{8}$ in. in diameter and weighing 3.5 g sinks 1 in. through a disk of bitumen held in a brass ring. The heating rate is 5°C per minute. The method of determining softening point is found in ASTM D36.

Thermoplastic materials, including the coal tars and asphalts, do not have a melting point, but soften over a range of temperature. A melting point is the temperature at which the atoms of a solid material lose their ordered crystal structure; since thermoplastic materials do not crystallize, they have no melting point. They are in a technical sense not solids but supercooled liquids, a statement that is also true of glass.

Bitumens applied as waterproofing below grade are never subject to high temperatures and have a low softening point, in the range 115–145°F, and a high penetration value. Bitumens used on roofs or vertical surfaces may be subject to radiant heat from direct sunlight, and require higher softening points and lower penetration.

The ductility of bituminous materials is important in many applications, including highway construction and roofing. Ductility is measured in a variant of the standard tensile test, with the specimen held at 25°C while it is elongated at a rate of 5 cm/min. Results are expressed as the total amount of stretch measured after fracture. The ductility of bitumens may vary over a range from zero to more than 150 cm, depending on the grade of material.

With prolonged exposure to weather and direct sunlight, asphalts, like some thermoplastics, such as polyethylene and polypropylene, slowly oxidize and lose their ductility. When such oxidized asphalts are strained, they can respond only by developing cracks. Overheating of bitumens during application also promotes their oxidation and embrittlement, and causes some loss of adhesive strength.

## 14.8 types of asphalt

While the petroleum asphalts are long-chain hydrocarbons, the coal-tar pitches contain considerable aromatics or ring compounds. Petroleum asphalts and coal-tar pitches are somewhat incompatible and cannot be used in combination with each other. If a layer of asphalt is applied over pitch, the asphalt will dissolve the more volatile ingredients in the pitch, producing a soft asphalt and a harder pitch.

Coal tars and pitches are used to saturate roofing felts and for the coating of kraft paper. Since these coal-tar materials soften at relatively low temperatures, they are used as roofing materials only on roofs with slight slopes or on flat roofs. Coal-tar bitumens oxidize more rapidly under exposure to ultraviolet radiation and must be protected by a coating of gravel or other aggregate. They have excellent cold-flow properties that make them self-healing, thus closing cracks that develop at low temperatures.

Frequently, asphalts must be oxidized to obtain desired properties, such as a required viscosity. Such asphalts are referred to as *airblown*. The chief use for oxidized (airblown) asphalts is in materials for built-up roofs. Blown asphalt is harder, less sticky, and more rubbery. It does not soften as readily on heating nor become as brittle on cooling as the nonblown material.

Asphalts are produced in three types:

1. *Hot asphalts:* softened by heating and applied hot.
2. *Cutback asphalts:* dissolved in solvents such as petroleum oils or naphtha.
3. *Emulsion asphalts:* emulsions of small droplets of asphalt dispersed in a water base.

The hot asphalts bond poorly to damp surfaces. They oxidize more rapidly when exposed to ultraviolet radiation and tend to be more brittle at low temperatures than the two cold types.

Cutbacks also bond poorly to damp surfaces; the emulsions bond well to such surfaces. The emulsions are a convenience for application, since the viscosity can be reduced simply by adding water. They dry by evaporation of the water in the emulsion. The emulsion types weather better than cutback or hot asphalts.

Paving asphalts are usually cutbacks. These are grouped into three general types:

1. *Rapid curing (RC):* uses gasoline as solvent.
2. *Medium curing (MC):* uses kerosene as solvent.
3. *Slow curing (SC):* uses heavy fuel oils as solvent.

The RC and MC types are the most used. In each type six viscosity grades are specified: RC-0, RC-1, RC-2, RC-3, RC-4, and RC-5. The higher numbers indicate higher viscosities.

The RC cutbacks have a penetration of 85–120; the MC types a penetration of 120–300. The viscosity of each cutback depends chiefly on the amount and kind of solvent used, since the curing rate is the rate at which the solvent is evaporated. Rapid-curing grades are used as binders for open-graded aggregates that coat quickly during mixing, or for surface treatments. Medium-curing grades are required for dense-graded aggregates, which need a longer mixing period.

Asphalt pavements use an aggregate bonded with asphalt, the asphalt functioning as a cement. Asphalts adhere well to basic (alkali) types of aggregate, such as limestone, but less tenaciously to acid types of rock, such as quartz.

The thermosetting plastics, after they are cured or hardened, cannot be resoftened by heating. They can be used at temperatures about 100°F higher than the maximum operating temperatures possible for the thermoplastics. This difference between the two classes of polymer materials results from a basic difference in the structure of the long carbon chains in each case.

Like the thermoplastics, the thermosets are composed of long-chain molecules, but in addition the chains are cross-linked, or cross-connected, to each other to produce a three-dimensional network. Figure 14.16 shows two molecules of phenol-formaldehyde cross-linked with carbon atoms. In rubbers, which are almost always thermosetting, this cross-linking is called *vulcanization*, the cross-linking agent being sulfur. The hardness, stiffness, and brittleness of these thermosets is increased by increasing the number of cross-links.

The practical differences between thermoplastic and thermosetting materials is best explained by a familiar comparison from paint technology.

Fig. 14-16   Three examples of cross-linking in thermosetting materials. Cross-linking requires a double bond, one bond to each molecule cross-linked.

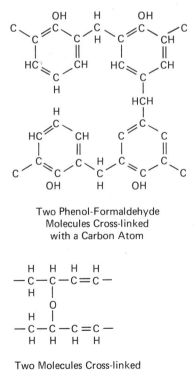

Two Phenol-Formaldehyde
Molecules Cross-linked
with a Carbon Atom

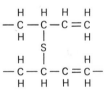

Two Rubber Molecules
Cross-linked (Vulcanized)
with Sulfur

Two Molecules Cross-linked
with Oxygen

Compare shellac, which is thermoplastic, and varnish, which is thermosetting. Shellac hardens by evaporation of its methyl alcohol solvent. Solvent loss is the only method of hardening a thermoplastic finish. Varnish, linseed oil, and the other thermosetting films harden by cross-linking. In the case of both varnish and linseed oil, the cross-linking agent is oxygen from the air. In the terminology of the paint industry, those finishes which harden by solvent loss are called *nonconvertible* finishes, and those that cure or cross-link are called *convertible* finishes. The most durable finishes are those that cross-link; similarly the best adhesives are also those that cross-link.

Wood is the outstanding thermosetting material used in the construction industry. It is superior in most properties to chemically manufactured thermosetting resins, although it absorbs moisture and is subject to fungus attack. The monomer of wood is a type of sugar, glucose $C_6H_{10}O_5$, shown in Fig. 12.3. The principal weakness of such a chemical is the presence of three OH groups: these attract water, which is a similar chemical, HOH, and are one reason for the absorption of water by wood.

The most important of the manufactured thermosets in the construction industry are the polyesters, epoxies, polyurethanes, and the various formaldehyde resins. A few others of specialized application must also be discussed under their applications. The rubbers are also thermosetting, but are not plastic materials. The polyurethanes, discussed in Chapter 15, are also used as finishes and sealants for buildings. Solid polyurethane rubber is of marginal interest to the construction industry, although widely used in other areas. In the following sections only the polyesters and the formaldehydes receive extended discussion.

## 14.10 formaldehyde resins

The several *formaldehyde plastics* (which are elastic) are familiar in telephones, electric switches and circuit breakers, television and appliance knobs, toilet seats, and melamine dinnerware. These thermosetting plastics are hard, strong, resistant to elevated temperatures, and brittle. They are entirely elastic under the conditions of the standard compression or tension test.

*Phenol–formaldehyde* is one of the lowest-priced members of the formaldehyde group. It is commonly called by its trade name, Bakelite. It can be supplied only in dark colors, therefore is used only in black telephone sets. This thermoset is used as the adhesive for exterior-grade plywood and is easily recognized by its dark brown color. About 4 per cent of the weight of a sheet of plywood is represented by the thermosetting adhesive.

A plywood bonded with phenol–formaldehyde will not delaminate under conditions of severe exposure to weather and water. This resin, however, has considerable color and sometimes will stain the plies. For decorative interior grades of plywood, urea–formaldehyde is preferred. For particle board, either urea–formaldehyde or phenol–formaldehyde is used as adhesive.

The *melamine–formaldehyde* and *urea–formaldehyde* resins have the same uses. Both may be colored. Melamine–formaldehyde has the better heat resistance and is self-extinguishing in flame tests. It is the hardest of the plastic materials and therefore the most scratch-resistant.

The following are approximate properties of the several formaldehyde resins. These properties may be altered by fillers and reinforcing materials.

| | |
|---|---|
| specific gravity | 1.5 |
| ultimate tensile strength | 8000 psi |
| elongation | 0.5% |
| modulus of elasticity | $1 \times 10^6$ |
| impact resistance | poor |
| thermal expansion | 0.0002 in./in./°F |

## 14.11 reinforced fiberglass plastics

*Polyesters* are the usual plastic materials used in fiberglass-reinforced structures. These are a group of plastics made from a range of chemical materials and include the Mylars, which are made into film and drafting paper, and the alkyds, used for wood and metal paints and enamels. Both saturated and unsaturated polyesters are used by industry. The *saturated polyesters* are cross-linked with styrene and thus are thermosetting; these are preferred for reinforced plastics.

The *reinforced polyesters* are now familiar in boats, hard hats, septic tanks, structural panels, corrugated roof panels, and a wide range of other structural applications, where their advantages of color, strength, and light weight can be exploited. The usual reinforcing is a matt of fiberglass. Ultimate strengths as high as 50,000 psi and $E$ values as high as $3 \times 10^6$ are common, and higher strength levels are possible. The glass fiber has a tensile strength of 150,000 psi, and the strength of the composite material is proportional to the amount of glass reinforcement used. The two materials, the polyester and the fiber mat, each by themselves offer little by way of structural properties; it is the combination of the two that produces useful properties. In a similar manner, a sandwich construction can combine two weak materials to make a strong composite.

A stress–strain curve for these reinforced plastics is very nearly a straight line to failure. This normally is the characteristic of a brittle material. But these are not brittle materials. Brittle materials will shatter on impact, while reinforced plastics are impact-resistant. Brittle materials are characterized by high compressive and low tensile strengths, but in the case of reinforced plastics these two types of strength are of the same order of magnitude. However, elongation is limited in a reinforced plastic.

Reinforced plastics will, like most other plastics, creep under sufficiently high long-term loading. In structural applications, a shape is preferred that will put the reinforced plastic in tension. Compression loads are best carried by a reinforced plastic incorporated into a sandwich construction.

## 14.12 hardness of plastics

## and elastomers

Hardness is related closely to strength, stiffness, scratch resistance, wear resistance, and brittleness. The opposite characteristic, softness, is associated with ductility. Hardness is a requirement of exterior building panels, whereas softness is necessary in caulking compounds and sealants.

Hardness is almost always measured by an indentation test. The Brinell, Rockwell, and Vickers indentation tests are used for metals. For asphalts the needle-penetration test is employed. The Shore durometer test, which also uses a penetrating needle, measures the hardness of plastics and rubbers. The Rockwell and Brinell testing instruments may also be used for hardness tests on plastics, but they are not suitable for rubbers. Typical Brinell hardnes numbers for plastics are the following, using a 10-mm steel ball under a 500-kg load for 30 sec:

| | |
|---|---|
| polystyrene | 25 |
| acrylic | 20 |
| PVC | 20 |
| polyethylene | 2 |

When using the Rockwell tester for plastics, the M and R scales are used. The M scale has a $\frac{1}{4}$-in. steel ball for an indentor preloaded with 10 kg, and measures the additional penetration from a load of 100 kg. The hardness is read from the red Rockwell scale of numbers. The M scale is used for the harder plastics. Softer plastics may be tested with the Rockwell R scale, using a $\frac{1}{2}$-in. ball and loads of 10 and 60 kg.

The hardness of a plastic as measured by the Brinell and Rockwell methods may be influenced by the thickness of the specimen.

An approximate comparison of Brinell and Rockwell R, M, B, and C scales is given in Fig. 14.17. Typical hardness measurements for plastic materials are these:

Fig. 14-17 Comparative hardness scales.

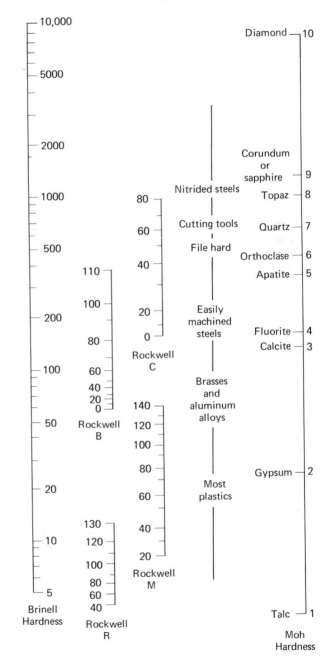

| melamine–formaldehyde | $R_M120$ |
| phenol–formaldehyde | $R_M115$ |
| urea–formaldehyde | $R_M115$ |
| polyester unreinforced | $R_M70$–110 |
| epoxy | $R_M80$–110 |
| acrylics | $R_M80$–100 |
| polycarbonate | $R_M75$ |
| polystyrene | $R_M75$ |
| ABS | $R_R105$ |
| nylon | $R_R110$ |

The *Shore durometer* for reading the hardness of plastics and rubbers (Fig. 14.18) is available in several ranges. The type A durometer is used for elastomers and soft plastics, the type D for harder plastics. Figure 14.18

Fig. 14-18   Shore durometer type A.

shows the type A instrument. The shape of the two durometer indentors is shown in Fig. 14.19. Both have the same diameter but different shapes of point. Both are pressed against the surface of the polymer against the resistance of a spring load, 822 g maximum for the type A instrument and 10 lb for the type D.

A simpler method of testing for the hardness of plastics is the *pencil test*. Actually this is a test of scratch resistance, similar to the Moh scale of hardness for rocks. The test uses the whole range of pencil hardnesses: from 2B, B, HB, F, H, 2H up to the maximum hardness obtainable, 9H. Each pencil is sharpened to a wedge and dressed with abrasive paper. The pencil is drawn across the surface of the plastic as it is held at right angles to the surface. If the plastic is harder than the pencil carbon, it will not be scratched. If the pencil is harder than the plastic it will groove the plastic slightly. Pencils are applied in increasing hardness until a hardness is found which first leaves a visible mark in the surface. The hardness of the plastic is reported as the pencil hardness grade. The pencil hardness method is not suited to elastomers.

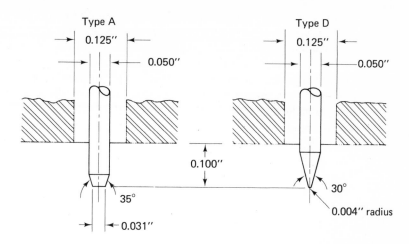

Fig. 14-19   Type A and type D indentors for the shore durometer.

## 14.13   elastomers

*Elastomers,* or *rubbers,* are capable of extreme elastic deformation at low levels of stress. In an extreme case a rubber formulation may be extended to 10 times its original length, although five times is more usual. The strain is not proportional to stress, as may be seen in the typical stress-elongation diagram of Fig. 14.3(d). This is a typical S curve exhibited by rubbers. In addition, there is a definite time lag, sometimes many minutes long, between stress and strain. This may be observed by indenting a rubber and watching the indentation showly disappear. This characteristic resembles the viscosity of stiff liquids. Unlike the metals and plastics, there is no change in volume when a rubber is strained. Rubbers therefore behave remarkably like elastic viscous

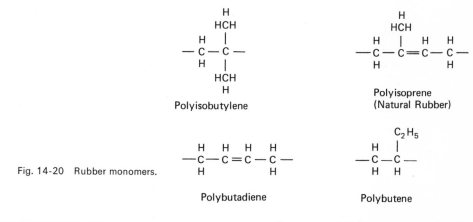

Fig. 14-20   Rubber monomers.

liquids; and, like liquids, most rubbers are not crystalline. Most rubbers are thermosets, the more familiar ones being cross-linked (vulcanized) with sulfur.

Rubber monomers are olefins unsaturated by hydrogen atoms and hence contain double bonds between some carbon atoms. Some of these rubber monomers are given in Fig. 14.20. When polymerized, these double-bonded monomers are susceptible to ozone attack, which results in crazing, and additives must be used in the rubber to protect the elastomer. Hypalon, having no double bonds, is impervious to ozone attack and thus can safely be used as a roofing surface. Unsaturated rubbers are also deteriorated by ultraviolet radiation, from which they are protected by substantial additions of carbon black.

There are so many kinds of rubbers employed as construction materials that a complete listing of them would be confusing. The many rubbers probably are better understood by dealing with them in their several applications as sealants, adhesives, finishes, roofing, and so on, giving only a few general remarks here.

The rubbers of interest to the construction industry are synthetic rubbers. Natural rubber is used in the tires of large earth-moving vehicles but not for standard car and truck tires, which are styrene–butadiene rubber (SBR). SBR is the standard or general-purpose rubber and is found in tires, conveyor belts, floor tile, sealants, and a multitude of other uses. The more important construction rubbers are these:

1. *SBR:* general-purpose rubber, used in cements, sealants, and floor tile.
2. *Butyl rubber:* copolymer of isobutylene with a small amount of butadiene or isoprene. This rubber, which is impervious to gases, is used for such products as tire inner tubes and the hoses through which the resins of sprayed polyurethane are pumped (if oxygen were to permeate through the hose, the polyurethane resins could solidify in the hose). Butyl is resistant to all the agents of outdoor weathering and is employed as a roofing membrane over sprayed polyurethane.
3. *Polychloroprene (neoprene):* used as electrical insulation and in certain types of curtain-wall sealants.
4. *Polysulfide (Thiokol) and polyurethane rubbers:* although chemically unlike, have similar applications. Both are used as rocket fuels and as curtain-wall sealants, and are better discussed under the topic of sealants in Chapter 16. Solid polyurethane is certainly a rubber, but polyurethane foams have, in addition, some of the characteristics of a plastic.
5. *Reclaim rubber:* reclaimed rubber of suitable quality is used in rubber cements.
6. *Silicone rubbers:* these are used as water-repellent treatments in liquid form on masonry materials, including concrete, mortar, brick, tile, stucco, and concrete paints. The silicone treatment prevents the penetration of water but allows internal moisture to pass out as vapor. Such a material is an ideal vapor barrier. The masonry material must not be silicon-treated

until it is well cured. As a solid rubber, silicones are quite soft and remark-ably heat resistant. They do not stiffen at low temperatures as do other rubbers.

7. *Chlorosulfonated polyethylene* (*Hypalon*): one of the few thermoplastic rubbers, although it develops some cross-links. It is applied by spraying to produce a roofing membrane in any desired color. Other applications of this very beautiful material will be discussed in later sections.

## PERTINENT ASTM SPECIFICATIONS

D2093-66T (Part 16) Preparation of Plastics Prior to Adhesive Bonding
E328-67T (Parts 30, 31) Stress Relaxation Tests for Materials
D2240-64T (Part 28) Indentation Hardness of Rubber and Plastics
D5-65 (Part 11) Penetration of Bituminous Materials
D36-66T (Part 11) Softening Point of Asphalts
D113-68 (Part 11) Ductility of Bituminous Materials
D1138-52 (Part 11) Resistance to Plastic Flow of Bituminous Mixtures

## INVESTIGATIONS

1 Investigate the pencil hardness test for plastics that is described in Section 14.12.

2 Make the flame spread test of Section 14.2 and report your results. Include some woods for comparison purposes. Always report the thickness of the mate-rial, since this has an influence on the flame spread rate. Record the charac-teristics of the flame, whether burning pieces separate from the specimen, and so on.

3 Differentiate between PVC and polystyrene by means of an ignition test.

4 Blow up a long balloon to about 9–10 in. in diameter. Spray a thickness of $\frac{1}{2} - \frac{3}{4}$ in. of polyurethane foam over the balloon in as uniform a thickness as

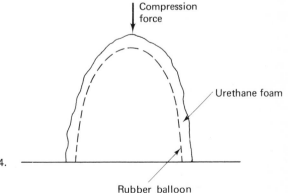

Fig. 14-21  Question 4, Chapter 14.

circumstances permit. The small aerosol cans of polyurethane are suitable for this purpose if a spray gun is not available. Cut off a length of the foamed balloon and load it in compression on a testing machine until failure (see Fig. 14.21).

The failure load should be impressive. Is the failure in compression or in tension? Brittle or ductile? Can you separate the rubber of the balloon from the foam? Can you account for what happened? At least one company in the United States produces such air-inflated structures.

5 Make a standard tension test of a round bar of polycarbonate, plotting the stress–strain curve. Measure the elongation after failure.

## QUESTIONS

1 The construction industry takes a conservative attitude toward new types of building materials—necessarily so. The following description applies to a standard building material, but it is written up as though it describes a new material about to be introduced to the industry. What is the material described? Would architects be sufficiently enthused by this description to try this material in their designs if it really was a new material?

"The material is an organic polymer. It can be fabricated readily into a variety of shapes that resist cold flow, and has considerable strength and stiffness. It is, however, subject to attack by various organisms, and loses strength by absorbing water. Its dimensional stability is not outstanding if subject to changes in moisture absorption, and it is readily ignited."

2 When the point of a drafting compass is pushed into a block of thermoplastic material, a small burr is raised around the hole. If the material is an elastomer, no burr is raised. Explain.

3 Classify the following as thermosetting or thermoplastic: (a) phenol–formaldehyde; (b) silicone rubber; (c) PVC; (d) polyethylene; (e) paper; (f) Mylar drafting paper; (g) foamed polystyrene; (h) foamed polyurethane (rigid); (i) most paints and varnishes; (j) most rubbers; (k) asphalt.

4 Define (a) a paraffin; (b) an olefin; (c) a polymer.

5 What characteristics of polyvinyl chloride make it a suitable material for a rain gutter?

6 Why does the softening point of a roofing asphalt have to be adjusted to suit the slope of the roof?

7 The application of asphalt over coal tar results in a change in the characteristics of the two materials. How does this happen?

8 Differentiate among hot, cutback, and emulsion asphalts.

9 What is the meaning of creep as applied to materials?

10 Account for the resistance of a saturated rubber such as Hypalon to ozone attack.

11  (a) Which rubber is a general-purpose rubber? (b) Which rubber is used in the carcasses of automobile tires? (c) which rubber is selected for impermeability to gases? (d) which rubber is used as electrical insulation?

12  You are offered a liquid vinyl spray that can be used as a roof coating to repair built-up roofs. The technical literature gives a tensile strength of 600 psi and an ultimate elongation of 250 per cent. Is this a plasticized or an unplasticized polyvinyl chloride? How do you know?

13  Acetal, a high-strength thermoplastic, has a polymer chain composed of alternating carbon and oxygen atoms. Hydrogen atoms are attached to the carbon atoms. Sketch the monomer of acetal.

A remarkable number of building materials are either cellular or expanded during manufacture. These include wood, a natural cellular material, perlite, vermiculite, and haydite, three ceramic materials that are expanded by heating processes, cellular glass, and foamed concrete. There remain to be discussed the foamed plastics, of which only five are extensively used in construction: polyvinyl chloride, polystyrene, polyurethane, polyethylene, and urea–formaldehyde.

Polyurethane foams are made in many types, flexible and rigid. The flexible polyurethane foams are foam rubbers, extensively used in furniture but rarely in building construction. Only the rigid insulating polyurethane foams are discussed in this chapter.

The only flexible foam used as a building material is *polyethylene foam*. It is a soft white foam with a somewhat feathery surface. Although not used in large volumes, it is an excellent flexible backup and is used also for expansion joints that will compress but not open up. It is bonded in place with asphalt.

# FOAMED
# PLASTICS

# 15

The *rigid foams* are extruded into various shapes chiefly as a substitute for wood. A cove molding of high-density rigid PVC is shown in Fig. 15.1. For building purposes, PVC is preferred and is normally used, since this material does not burn. In finished form such extruded shapes may be dif-

Fig. 15-1   Foamed plastics. A polystyrene beadboard on the bottom, two sections of high-density foamed PVC ceiling cove molding above it. The three materials at the top of the photograph are from left to right, foamed polyurethane, foamed polyethylene (feathery appearance) and foamed urea-formaldehyde (snowy appearance).

ficult to distinguish from wood when stained a dark brown. When molded into panels, even the open grain of hardwoods can be simulated. Rigid PVC is extruded into moldings, baseboard, window and door trim, fence posts, fence palings, shelves, pipe, siding, soffits, and door sills. Even the density of these foams corresponds to that of wood—about 30 pcf. Such extrusions have a lower modulus of elasticity than wood, however, and for window and door frames require sufficient cross section to give them stiffness. Only polyvinyl chloride high-density foams are used for exterior applications.

A characteristic of these rigid structural foams is a dense skin for about $\frac{1}{8}$ in. from the surface (Fig. 15.2). This skin provides added strength and durability.

These high-density foams are not suitable for use as insulating foams, although a limited amount of lightweight foamed polyvinyl chloride has been used as building insulation.

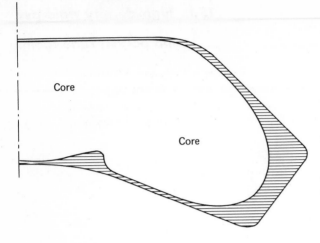

Fig. 15-2 Cross section of the PVC foamed cove molding of Fig. 14-5 showing the shape of the higher-density integral skin of this extrusion.

## 15.2   urea-formaldehyde

### insulating foam

*Urea–formaldehyde* is the lightest of the insulating foams, weighing only 0.7 pcf. Its $K$ factor is 0.2 Btu/ft²-hr-in.-°F. It has virtually no mechanical strength, is white in color, and does not support combustion.

The polystyrene and polyurethane insulating foams to be discussed next have closed cells and therefore do not absorb water. About 60 per cent of the cells of foamed urea–formaldehyde are open (connected), and this foam can absorb about 30 per cent water by volume.

Like polyurethane foam, urea–formaldehyde is foamed on the jobsite. The foaming system has two components, like polyurethane foam. One is a water solution of urea–formaldehyde and the other a water solution of the foaming agent and a catalyst for curing the resin. The sprayed foam from the spray gun has enough viscosity to remain in place on an open vertical wall. It does not, however, bond to surfaces as polyurethane foam does, but will flow into crevices and hollow spaces before setting up. Polyurethane foam will stick in place and not flow.

Of the insulating foams, polystyrene is thermoplastic, and urea–formaldehyde and polyurethane are thermosetting.

## 15.3   polystyrene insulating

### board

In the case of the foamed insulating plastics—polystyrene, polyurethane, and urea–formaldehyde—the volume of solid material is very small, about 2 pcf or less, and is used simply to enclose cells of gas or air. Therefore, the thermal conductivity of these materials is closely that of the gas enclosed in the cells. Air is not the best insulating gas; carbon dioxide and Freon gas are superior. The insulating gas in foamed polystyrene is air or pentane, which results in a *K* factor of 0.24 Btu for this material.

*Foamed polystyrene board* is available in two types, beadboard and expanded polystrene (styrofoam). The cells are larger in beadboard, and the material has a beaded appearance. Beadboard is produced by expanding polystyrene beads and bonding them by heat in a mold. The alternative method is to include a foaming agent in polystyrene resin and to extrude the material through an orifice as planks. The foaming agent for both processes is usually pentane, $C_5H_{12}$.

Various densities of foamed polystyrene board are produced. The higher densities have higher *K* factors. The coefficient of expansion is 0.00004/°F. Most construction grades are made of flame-retardant polystyrene.

Foamed polystyrene board has only limited resistance to heat and is attacked by most solvents, including those in many paints and adhesives. It cannot be attached with standard asphalts or solvent types of adhesive. Special coal-tar epoxies are formulated for bonding this material. The material is of course brittle and must be carefully handled. Although inferior to polyurethane foam or foam board in almost all properties, foamed polystyrene is cheaper in first cost.

Polystyrene foam planks are used as insulation, including perimeter insulation (Fig. 15.3), as form liners for concrete pours, as insulating lath for plaster, as roof insulation below the built-up roof, and as underfloor

Fig. 15-3  Perimeter and wall insulation for a warehouse building in a cold climate.

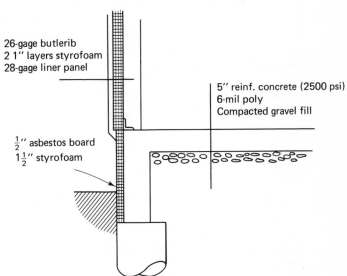

26-gage butlerib
2 1″ layers styrofoam
28-gage liner panel

5″ reinf. concrete (2500 psi)
6-mil poly
Compacted gravel fill

$\frac{1}{2}$″ asbestos board
$1\frac{1}{2}$″ styrofoam

insulation upon which concrete is poured. When used as a concrete form, the purpose usually is to bond the concrete and the insulation, the insulation remaining when the forms are removed. As a roof insulation, the foam should be encased in asphalt-laminated kraft paper to protect the foam from the hot asphalt. A heavier-density foam is preferred for roofing, because the heavier material is more resistant to collapse from the heat of hot-mopped asphalt and also more resistant to damage from ice formation underneath the built-up roofing.

## 15.4   rigid polyurethane

### insulating foam

*Polyurethane foam planks* are available in various densities: 2, 3, 4, 6, 8, and 10 lb per ft³. These are produced by foaming a very large volume of material, perhaps 4 × 4 ft in section and sawing this large "bun" into planks on a bandsaw. Such planks have the same applications as styrofoam planks. They are, however, not quite so brittle, are somewhat stronger, will withstand temperatures about 100°F higher than styrofoam can tolerate, and they are not attacked by solvents. The 2-lb density is usually installed, but for roofing insulation, a 4-lb density is more resistant to damage by ice formation in the roof. Polyurethane is not harmed by hot-mopped asphalts.

Polyurethane foams are produced by a reaction between a diisocyanate such as tolylene diisocyanate and a polyester. A small amount of water is used in the reaction in order to form carbon dioxide. This $CO_2$ is the foaming agent, although Freons may be the blowing agent in foams of low density. The use of Freon may provide a lower $K$ factor. A silicone oil or other surface-active agent keeps the cells very small. The polymerization reaction produces linkages of urea with ethane, hence the name polyurethane. The density of the foam is determined by the type of polyester used and the amount of water, about 3 per cent, which controls the amount of $CO_2$ generated.

The gas in the closed polyurethane cells is therefore either $CO_2$ or a Freon. The thermal conductivities (in Btu) of these gases and of air are these:

|  |  |
|---|---|
| Freon-11 | 0.058 |
| $CO_2$ | 0.102 |
| air | 0.168 |

Thus the Freon-blown polyurethanes are the best insulation. Both the $CO_2$ and Freon make polyurethane a superior insulating foam to styrofoam. However, any comparisons must take account of an aging process that may occur in polyurethane foam. Freon-foamed polyurethane in a density of 2 lb per ft³ has an initial $K$ factor of 0.11, far superior to any other insulation.

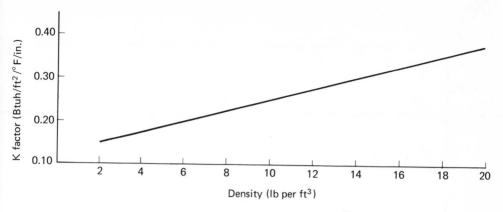

Fig. 15-4   *K*-factors for polyurethane foams.

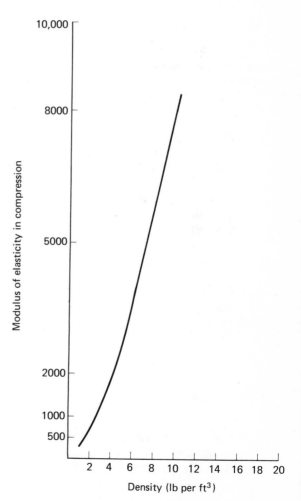

Fig. 15-5   Average values for modulus of elasticity of polyurethane foams. Specific types may vary somewhat from these values.

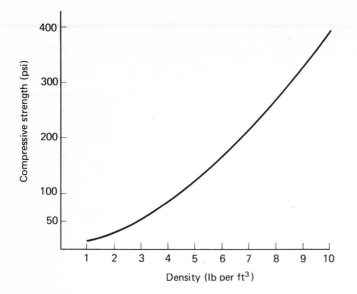

Fig. 15-6 Average values of compressive strength of polyurethane foams.

If, however, this insulation has access to air, air will permeate through the
cell walls and replace the Freon. This exchange of gases raises the $K$ factor
to 0.16 Btu, which is, of course, still superior to styrofoam. If the foam is
enclosed so that air does not have access to it, the $K$ factor should remain
at 0.11 Btu.

Unlike styrofoam, which has substantially uniform properties with
respect to orientation, polyurethane foams are stronger in the direction of
foam rise than in the cross-direction. Depending on which direction requires the better strength, sandwich panels may be foamed in the vertical or
the horizontal position.

Significant properties of polyurethane foams as a function of density
are given in Figs. 15.4 –15.6. It will be noted that mechanical properties
such as modulus of elasticity, tension, compression, and shear strength are
not proportional to density but increase at a more rapid rate.

## 15.5 installation

For minor installations on the jobsite, aerosol kits of the two polyurethane
resins, complete with nozzles for spraying, are supplied by at least two
manufacturers (Fig. 15.7). These kits are available in quantities to spray
as little as 9 board feet of insulation.

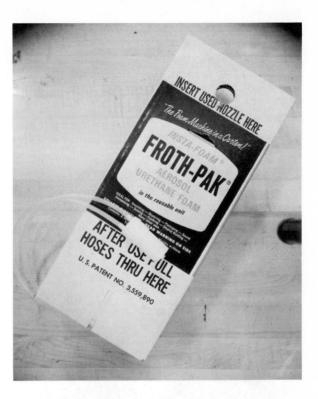

Fig. 15-7 Small aerosol kit containing two components for urethane foam complete with spray nozzle.

Fig. 15-8 Small mobile polyurethane spray unit. (Photo courtesy Gusmer Corporation.)

For standard spray contracting, such as insulating the walls and roof of a building, a complete spray installation, truck-mounted, is necessary (Fig. 15.8). The complete installation requires the spray gun, hose, resin heaters, proportioning pumps for the two components, and at least one 55-gal drum of each component. The two components are mixed in equal volumes for building applications, although not necessarily in equal volumes in the furniture and other industries. The contents of the two drums will cover 3000–4000 ft² to a 1-in. thickness. The sprayed liquid foams immediately upon being deposited on a surface and hardens in about 1 min. Installation of polyurethane by a spray system requires equipment of the highest quality, because the resins are very viscous and highly reactive. If the two resins react in the gun and set up, they cannot be removed by any solvent. A self-cleaning gun is almost always used (Fig. 15.9).

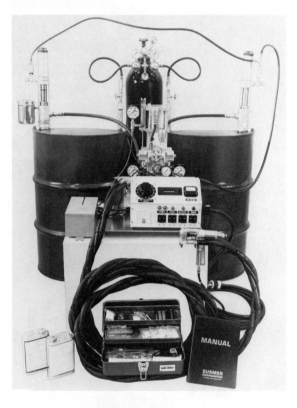

Fig. 15-9  Polyurethane spray unit. A pump is installed in each drum of resin. The nitrogen bottle between the drums provides nitrogen gas to protect the resin in the drums from atmospheric air and water vapor. The hose lines are of impermeable butyl rubber to protect the resin from air and water vapor. The orifice is on the left-hand side of the gun, the same direction as the ends of the hoses point to. (Photo courtesy Gusmer Corporation.)

It is not economical to spray polyurethane in small amounts, a single house for example, because of the lost time required to bring the rig to the site, set it up, adjust the gun, and clean up afterward.

Styrofoam and polyurethane foam compete with each other as insulation. Styrofoam boards cost about 10 cents per board foot at the time of writing; the resin cost of urethane foam is also about 10 cents per board foot without installation cost. On this basis styrofoam would rarely be used. Polyurethane is elastic, whereas styrofoam is brittle. Polyurethane foam is its own adhesive; styrofoam must be bonded in place. Polyurethane is a superior insulation and when sprayed can fit any contours and has neither joints nor gaps. However, in the simpler applications, such as perimeter insulation, the cost of spraying may exceed the cost of the polyurethane resins, but the cost of installation of styrofoam board would be relatively low. On a great many jobs however, the superiority of polyurethane warrants a cost increment, or in other cases may actually produce a cheaper installation.

Neither styrofoam nor urethane provides an effective vapor barrier. Both supply some degree of acoustic absorption. Both are degraded if exposed to sunlight.

## QUESTIONS

1 Figure 15.10 shows a shipping container, type MLW/MLLU 24, used in transatlantic shipping. This is a nonrefrigerated container with an allowable

Fig. 15-10 International Standards Organization (ISO) merchandise shipping container type MLW/MLLU 24, nonrefrigerated insulated, capacity 41,260 lb. The allowable heat leak from this container is 47 btu/hr°F.

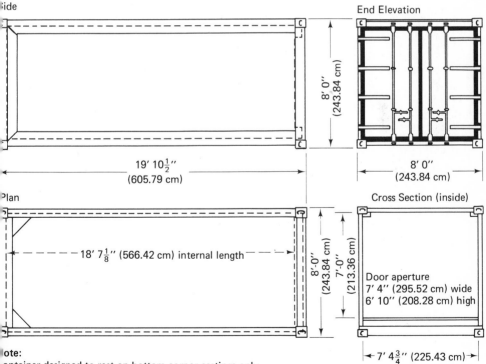

Note:
Container designed to rest on bottom corner castings only.
*Do not chock under floor.* Floor loading for fork-truck operations in container not to exceed 6000 lb on 22 in.$^2$ (2721.6 kg on 141.9 cm$^2$)

*foamed plastics* / 263

heat leak of 47 Btu/hr per degree of temperature difference over the whole area of the container. Container dimensions are 19 ft 10½ in. long × 8 ft 0 in. high × 8 ft 0 in. wide. The container is insulated with polyurethane foam with a design $K$ factor of 0.125 Btu. How much foam insulation is required to insulate the container?

2 Figure 15.3 is a detail of the perimeter insulation for an industrial warehouse. The ⅛-in. asbestos board contributes little to the heat insulation. Why is it specified?

3 What advantages does polyurethane foam offer over styrofoam?

4 Explain how the $K$ factor of polyurethane foam, 2-lb density, can change from 0.11 to 0.16.

5 What are the uses of high-density polyvinyl foam?

6 Under what circumstances would urea–formaldehyde insulating foam be preferred over other types of synthetic foams?

Sealants must be distinguished from sealers. *Sealers* are coating compounds used to seal the surface of materials against penetration by water. *Sealants* are materials used to seal joints in buildings or pavements by expanding and contracting with the movement of the joint. Such joints may be expansion joints, joints between glazing sheets and their frames, or the joints in curtain-wall construction. Caulks or *caulking compounds* are less extensible materials than sealants and are used in less critical applications. Caulking compounds are not expected to be stressed in tension. Glazing compounds are basically cushioning materials to retain glass in frames so that movement in the frame is not transmitted to the glass.

## 16.1   performance of a sealant

A sealant must maintain a closed joint under any conditions of expansion, contraction, or other reasonable movement of the two parts of the structure that it is intended to seal. The sealant must also serve as a deformable cushion if it seals a panel of glass, because if glass is subject to simultaneous flexural and thermal stresses, it is likely to crack.

# SEALANTS
# AND
# SEALERS

# 16

Suppose that the joint width extends as the result of a temperature change. The sealant must deform considerably to keep the joint closed. Later, under different temperature conditions, this strain will be relaxed, and the sealant will be expected to return to its original dimension and shape. Clearly the sealant must be elastic and not plastic in its stress–strain behavior; a material such as polyethylene will not do.

A material with a high modulus of elasticity could not be used as a sealant. Construction joints quite often enlarge by 100 per cent or more. A strain of 100 per cent would mean a stress in the sealant equal to the modulus of elasticity (assuming a linear stress–strain relationship). No material is capable of a stress as large as its modulus of elasticity. Again, it is clear that sealants must be made of materials of low modulus, which is of course the reason that sealant materials are soft.

Because sealants are required to be capable of very large elastic strains, elastomeric materials are ideally suitable materials. In practice, a few other materials are used which are more limited in their capacity for elastic strain, such as vinyls and asphalts. These materials may be used in joints that have very limited movement. The peculiar S shape of the stress–strain curve of elastomers, of which Fig. 14.3 is an example, is ideal for a sealant. The initial resistance of the elastomer to strain quickly falls to a very low level, permitting the material to elongate remarkably without developing large stresses.

## 16.2   sealant methods

Sealants are applied in any of three methods:

1. Putty-like mastic.
2. Tapes, ribbons, and beads.
3. Preformed gaskets.

The *mastic sealants* must have a controlled viscosity to suit the requirements of the application. Those applied to horizontal joints should be self-leveling; for vertical joints they must not sag. Mastic sealants are employed in one- and two-component formulations. The traditional sealants and caulking compounds such as asphalt and putty are examples of one-component systems. These require no mixing before application. They cure by drying or by oxidation. Two-component systems have two substances that are mixed together and react with each other (cure). Often the two components are designated as parts A and B. The B component is usually the curing agent or catalyst, with an accelerator sometimes added.

Mastic sealants may be applied either with a gun that extrudes the sealant from a cartridge container, or with a knife such as a putty knife.

Those formulated for application with a knife have a stiffer consistency than the gun grades.

*Continuous tapes and beads* are supplied in sheets or rolls for pressing into place. These contain structural fillers to provide cohesive strength.

*Preformed gaskets* are prepared as rings, tubes, or other shapes which are inserted into the joint. These have only moderate elongation. The water-stop of polyvinyl chloride shown in Fig. 16.1 is an example of a preformed resilient gasket.

Fig. 16-1   Polyvinyl chloride waterstop. The serrations provide a long leakage path to impede the flow of water. This waterstop is suitable only for compressive deformation.

Joints in highway pavements may be filled with hot-poured asphalt. Consistency is tested by ASTM D-5, which measures the penetration of a cone into the asphalt under 150 g. A bond test for adhesion to asphalt is another of the more important tests applied. For this test the asphalt is poured 1 in. thick between two $1 \times 2 \times 3$ in. concrete blocks, and a tension pull applied.

## 16.3   design of a sealant
## for high elongation

If a mastic sealant must extend in order to keep a joint closed, it must bond successfully to the sides of the joint. When this bond fails, the sealant has failed. Such a bond would not be necessary in the case of a precompressed resilient sealant.

The most important characteristics determining the successful functioning of a mastic joint sealant are, besides the stress–strain properties of the sealant material, the width of the joint, the total movement of the joint, and the thickness of the sealant.

Suppose that the installed gap in the joint is $\frac{1}{8}$ in. If the maximum movement expected in the joint is $\frac{3}{8}$ in. the sealant must deform 300 per cent. No existing sealant is capable of such deformation over an extended period of time, even though most elastomers are capable of this deformation in a laboratory stress–strain test. If the gap in the joint is made $\frac{3}{4}$ in., and the maximum movement expected is $\frac{3}{8}$ in., an elongation of 50 per cent is required of the sealant. Such an elongation can be provided over a period of years by the superior grades of sealants. Therefore, the basis of joint design is

$$\frac{\text{elongation}}{\text{initial gap}}$$

and this ratio must not exceed what the sealant can deliver over the period of its useful life. These considerations explain the large joints often found in curtain-wall and pavement construction. The greater the expected movement, the larger must be the joint gap.

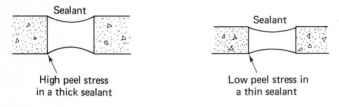

High peel stress in a thick sealant

Low peel stress in a thin sealant

Fig. 16-2   Necking effect in sealants.

The thickness of the sealant deposited in the joint must not be large. The deeper the sealant depth, the more necking and internal stress occur in the sealant (see Fig. 16.2). The deeper joint risks a necking failure in the sealant. Therefore, a backup material is used to fill the greater part of the joint, and the sealant is installed as a topping layer, as in Fig. 16.3. The sealant must not adhere to the backup material, since such a bond will restrain the elongation of the sealant. Polyethylene or masking tapes are often used to separate the mastic from the backup material. The depth of the sealant should be $\frac{1}{8}$ to $\frac{1}{4}$ in. but not greater than its width.

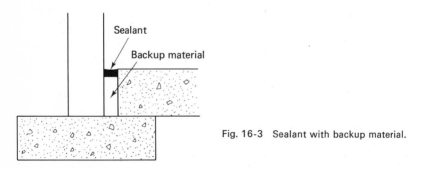

Fig. 16-3   Sealant with backup material.

If the joint is in a traffic-bearing floor, the backup material must be capable of supporting the sealant against traffic pressures. Rigid foams of rubber or plastic, cork, fiberboard, solid rubber, and preformed gaskets are the materials used as supporting backups.

For glazing aluminum sash in curtain walls, a minimum sealing gap of $\frac{1}{8}$ in. is required. The lap or bite (Fig. 16.4) is specified by the manufacturer but is usually $\frac{1}{4}$ in. for glass less than 50 united inches, $\frac{3}{8}$ in. for 50–100 united inches, and $\frac{1}{2}$ in. for a glazing area exceeding 100 united inches. The glass is set on setting material laid on the sill near each corner and centered

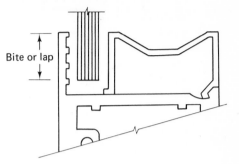

Bite or lap

Fig. 16-4   Bite or lap of a sealant.

by shims on each side. The glass must not rest on the metal sash. A minimum installation temperature of 40°F is required, since any condensation will prevent the mastic from bonding.

## 16.4   sealant materials

The standard *linseed oil putty* is used only for glazing wooden sash. It is compounded of 90 per cent fine calcium carbonate (whiting) and 10 per cent raw linseed oil. Some types also include white lead. It sets slowly by oxidation of the linseed oil, and in time becomes quite hard and brittle. Its effective elongation therefore is zero, and its adhesion is doubtful.

Asphalts, like glazing putty, harden slowly by oxidation. Some of the asphalt sealants include fiber for stiffening. Other types of sealants are compounded of oils and solvents, or elastomers such as butyl rubber, and harden initially by solvent release. Such sealants do not adhere well, and the loss of solvent results in some shrinkage. Permanently plastic caulking compounds are based on high-viscosity oils such as soybean, cottonseed, or on synthetic elastomers such as polybutene or polyisobutylene. These frequently contain clays or asbestos or other fiber to improve cohesion and to ensure that the caulk does not sag. These permanently plastic formulations have no elastic recovery, and therefore may be used only to seal joints that are not subject to thermal movement or vibration. They are, however, used as cushions and backup materials in joints.

The *load-bearing* or *stressed sealants* require a chemically cured elastomeric binder. The binders currently in use include polysulfides, silicones, urethanes (not the types used as foamed insulation), polymercaptans, chlorosulfonated polyethylene (Hypalon), butyl, polybutene, polyisobutylene, acrylics (not the types used for glazing), and polychloroprene (neoprene). All formulations, like all materials everywhere, represent a compromise of properties. No one sealant has the highest and most permanent elastic properties or the best adhesion.

The *polysulfide rubbers* (Thiokol) were first used in World War II as a self-sealing linear for aircraft fuel tanks perforated by shrapnel or machine gun bullets. They were the first of the new polymeric sealants to be applied to curtain-wall construction. The polymer requires two components, the liquid polymer being part A, with a part B oxidizing curing agent. One-component systems are also available but may have long curing times. A polysulfide sealant will give effective elongations of almost 100 per cent over long periods of time, and does not become brittle at low winter temperatures. After some years, a polysulfide sealant will reach a terminal hardness of about Shore A 70, which is the hardness of an automobile tire.

The *acrylic sealants* are one-component types, with good elongation and adhesion, but they are somewhat expensive.

*Silicones* are polymers with a silicon–oxygen chain with attached carbon side chains. The silicone rubbers are longer polymer chains than the liquid silicones used to waterproof masonry, and are cross-linked (vulcanized). They are often referred to as RTV rubbers (room-temperature vulcanizing). The polymer is clear and is there fore easy to color with pigments, although only a restricted range of standard colors is offered to the construction industry. Since the silicone rubbers are not polymerized from unsaturated olefins as most rubbers are, but from a saturated monomer such as Hypalon, they are not degraded by ozone, oxygen, or ultraviolet radiation. They have high elastic recovery, are quite soft, and do not become brittle at low temperatures. They are, however, more expensive than most sealants. Silicone sealants have a consistency similar to toothpaste when extruded from the applicator gun. They may be extruded into place at any temperature and do not sag at high temperatures. They cure by combining with atmospheric moisture.

*Urethane sealants* are available in one- and two-component systems. One-component urethanes cure by combining with atmospheric moisture. The urethane sealants offer the advantages of oil resistance, flexibility at low temperatures, high elastic recovery, and resistance to abrasion which exceeds that of any other sealant. Their low-temperature flexibility is exceeded only by that of the silicones. If elongated 50 per cent for a full year under outdoor exposure conditions, a urethane sealant will recover completely to its original length. However, these excellent properties are degraded by exposure to ultraviolet radiation from the sun. The abrasion resistance of urethane sealants makes them well suited to floor and pavement joints subject to traffic loads.

*Chlorosulfonated polyethylene* or *Hypalon* is a thermoplastic or nonvulcanizing rubber with outstanding weather resistance. It can, however, be cross-linked with oxygen, although not with sulfur like other rubbers. Being a white rubber, and needing no carbon black for protection against ultraviolet radiation, it can be colored. It is abrasion-resistant, although it is not recommended for traffic conditions. Adhesion of Hypalon is good, although not outstanding. Hypalon loses resilience at temperatures below 0°F, however.

When used as a sealant, Hypalon is a one-component system which is normally provided in white or bright pastel colors, with excellent resistance to weather, water, and ozone. Both knife-grade and gun-grade viscosity are available. Considerable xylene solvent is used in the formulation of Hypalon, most of which evaporates, resulting in some shrinkage of the sealant.

*Butyl rubber* is a nearly saturated rubber, and since there are few double carbon bonds in the carbon chain, it is weather-resistant and thus suitable as an exterior sealant. The recovery of a butyl sealant is somewhat limited, but it is serviceable as a sealant at low temperatures. It does not harden very much with age. Any environmental attack depolymerizes and softens it. Butyl rubber is a solvent-release type of sealant. The cheaper SBR rubber (styrene–butadiene) is also used in solvent-release sealants for interior and less critical applications because it does not age well.

*Neoprene* (*chloroprene*) ages well and resists weathering, hence finds considerable use in sealants. It like Hapalon, stiffens at low temperatures. Neoprene sealants are usually colored black, being compounded with carbon black for ultraviolet protection. If a color is needed, a Hypalon coating is provided. Like all solvent-release types, neoprene sealants shrink on curing. Elastic recovery is somewhat limited.

### 16.5 gaskets and tapes

Any of the elastomers discussed above may be used as preformed resilient gaskets for sealing. The usual rubbers selected for such gaskets are butyl,

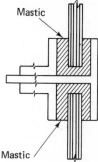

Fig. 16-5  Mastic sealant using a U-shaped installation.

polybutene, polyisobutylene, and neoprene. Certain plastics, such as plasticized polyvinyl chloride, may be used if not exposed to weather and if movement of the joint is limited.

For widow installations, U-shaped channels are often employed. These snap around the two sides of the glass (Fig. 16.5). A 25 per cent initial compression is recommended for glazing gaskets.

Sealing tapes are also used in window glazing and other applications (Fig. 16.6). These are preformed semisolid materials with some adhesion, with the advantage of convenience in application, since no special equip-

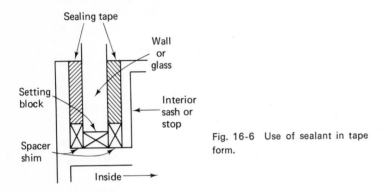

Fig. 16-6  Use of sealant in tape form.

ment is needed, and there is no delay for curing. The tall John Hancock Building in downtown Chicago used a tape-glazing system. Such tapes usually are made of polybutene, polyisobutylene, or butyl rubber. Polybutene is a saturated elastomer; the others have limited unsaturation.

## 16.6  sealers

No doubt all vapor-barrier materials—asphalt-laminated papers, polyethylene film, and others—might be considered to be sealers. Usually, however, a sealer is considered to be a liquid material that adheres to a surface to protect that surface from penetration by moisture.

Asphalt coating mixtures are extensively used as sealers to protect buildings, poles, and docks from water and rot (Fig. 16.7). *Damp-proofing* means the application of an unreinforced asphalt to a concrete or masonry surface, usually subgrade, which is subject to limited water exposure. *Waterproofing* is the use of a fiber-reinforced asphalt on such surfaces to prevent entry of moisture under conditions of hydrostatic pressure. Asphalt sealers for these purposes are applied either by brush or spray. For waterproofing, as many as three to six fiberglass reinforcing mats may be applied for adverse conditions of groundwater pressure.

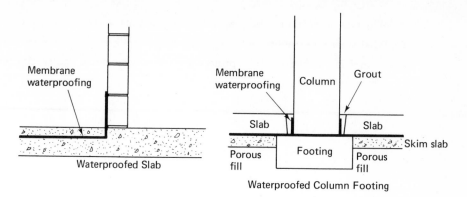

Fig. 16-7 Waterproofing.

Clear silicone sealers for moisture-proofing of masonry produce no change in the appearance of the masonry. Hypalon serves a similar purpose, providing a flexible, decorative, and protective sealing coat on metal, wood, or masonry surfaces.

Sealers for metal surfaces include vinyl coatings, coal-tar epoxy coats, Hypalon, chlorinated rubber, butyl rubber, and aluminum paint. Chlorinated rubber, produced by reacting rubber with chlorine, protects against many corrosive agents, as well as water. It may also be applied to masonry and concrete.

The *Elastron* butyl rubbers are excellent protection against moisture and chemical attack. They are applied as solvent-release liquids, usually by spraying. They will bond to virtually any surface, including wood, masonry, concrete, steel, urethane foam, and even polyethylene.

When applying any liquid membrane that hardens by solvent release, at least two coats are necessary. The second coat is needed to close any pinholes that may develop in the first coat due to evaporation of the solvent.

*Aluminum paint* is a dispersion of very fine aluminum flake in an asphalt. It has the excellent adhesion of asphalt to virtually any type of surface. The flakes of aluminum interleave to provide excellent resistance to penetration by water, even in thin coats.

## QUESTIONS

1 Differentiate between a sealer and a sealant.

2 What characteristics of the elastomers make them suited to sealant applications?

3 Explain why a wide joint makes less severe demands on a sealant.

**4** Why must a sealant not be bonded to its backup material?

**5** What is an RTV rubber?

**6** In applying a solvent-release protective coating, why would you specify two coats?

The components and separate pieces of a building must be connected together either by mechanical fasteners or adhesive materials. The mechanical fasteners such as nails, lag screws, and masonry anchors are construction items that have become familiar through the experience of many decades. Some of the adhesive materials, such as the hydraulic cements, mortars, and asphalts, are equally familar. A great many of the newer adhesives have been developed since World War II, and the number of these increases with the continuing development of polymer technology. The characteristics of these more recent and less-well-known adhesive materials must be discussed in this chapter.

Adhesives and surface coatings are related, and many materials are formulated for either purpose, such as the expoxy adhesive and epoxy paint materials, or acrylic adhesives and paints. In both cases a thin film must be attached to a substrate material. The adhesive must bond only. The surface coating must bond but has the additional functions of protecting the substrate and providing an attractive appearance.

The range of surface finishes is very wide and includes organic materials, such as linseed oil, varnishes, and enamels, and inorganic materials, such as porcelain enamels and anodized aluminum finishes. The thermosetting

# ADHESIVES
# AND SURFACE
# FINISHES

# 17

adhesives and finishes are generally superior to the thermoplastic types. The former do not soften at elevated temperatures, have higher strengths, and their cross-linked chemical structure makes them resistant to chemical attack and weather. Hence the use of thermosetting phenol-formaldehyde as an adhesive for exterior-grade plywood.

## 17.1  surface preparation

No coating or adhesive can attach to a surface unless it can "wet" the surface. An example of a nonwetting condition is a drop of mercury on a steel plate. The mercury has no attraction for the steel plate and will not spread on it; instead, it draws into a tight sphere that can be rolled over the plate. Mercury therefore cannot be used to bond steel to another material. A drop of a wetting material will spread over the surface of the plate, and if it can be made to harden or cure to a solid, will produce an adhesive bond or a surface coating. Whether a coating material can wet a surface depends also on the condition of the surface. No bond would be expected with an oil-covered or rust-covered steel plate.

Many construction methods involve wet processes, as in bricklaying, plastering, and concrete work. There is always pressure to complete the building for earliest occupancy. There is thus the temptation to spray urethane foam, or to lay asphalt roofing and other materials on concrete that is not completely cured. If an impervious film of paint is applied over materials containing moisture, this moisture is trapped and will soon break the bond between paint film and substrate material, with the production of blisters and peeling. Emulsion paints are a good choice for surfaces that are not quite ready to be given an impervious paint coat.

Paint systems formulated for wood and metal cannot be applied successfully on concrete. Oil paints usually fail because alkalis in the concrete react chemically with the oils. Cement paints and some types of emulsion paints are suitable for concrete.

When small amounts of special mortars or cements must be mixed, the applicator often prefers to mix such materials in a polyethylene pail. Adhesives and paints do not stick to polyethylene, and the pail is easily cleaned. However, polyethylene foam can be bonded with adhesives, and the explanation is of course that the adhesive can enter the pores of the foam and lock mechanically to the polyethylene. The roughening of a smooth surface by sandpapering with coarse sandpaper before applying adhesive follows the same principle. Similarly, a reasonable amount of suction aids in bonding mortar to masonry units. These mechanical aids to bonding assist in making a strong joint, but the primary requirement for adhesives and coatings is the wetting of the surface. For example, glass is very smooth; nevertheless paint adheres well to glass, because it wets the glass.

Movement between two bonded materials may break the joint. Such movements arise usually from changes in temperature or moisture content. To ensure that the bond holds, an elastic adhesive must be used for such conditions. For extremes of movement, an elastomeric sealant in a thick joint must be employed. It is usually desirable for the stiffness of an adhesive to be less than that of the materials to be bonded. If this is the case, the adhesive can absorb and relax stresses.

The use of asphalt as an adhesive has been mentioned repeatedly in this book. Even though it is thermoplastic and its chemical composition is unknown, asphalt wets and bonds to most of the construction materials. It even develops some bond strength to polyethylene. Its resistance to moisture and weathering is quite satisfactory. Sprayed polyurethane foam has the ability of asphalt to wet and adhere to most surfaces.

### 17.2 types of adhesives

The available range of adhesives is currently quite wide, the following list summarizing the range:

1. animal and vegetable glues
2. inorganic adhesives and cements
3. solvent-release thermoplastic adhesives
4. chemically curing thermosetting adhesives
5. elastomeric adhesives
6. emulsion adhesives
7. pressure-sensitive and contact adhesives
8. hot-melt adhesives

The *animal and vegetable glues* are used to bond paper, leather, wood, and cloth. This group of adhesives includes starch, casein, animal and fish glue, and many others. These adhesives are not reliable under conditions of elevated or low temperatures. They are usually liquefied by adding water, hence cannot be used under conditions of moisture penetration. With the exception of casein, these glues have limited use in construction. Starch pastes are sometimes used in paper- or cloth-backed vinyl wall coverings.

Many construction workers smoke cigarettes, and casein glue fills two functions for the cigarette smoker. This adhesive bonds the seam of the cigarette paper and also the strip of abrasive on the match-book cover. Both joints require water resistance.

Asphalt falls into two of the above groups of adhesives. An emulsion asphalt is an emulsion adhesive; a cutback asphalt is a solvent-release adhesive. Many resilient flooring applications employ the versatile asphalt cements as an emulsion, although sulfite flooring adhesives are also used above grade (linoleum paste). Asphalt, like the sulfite liquor adhesives, is resistant to the alkalis in cement and to water.

The *inorganic adhesives* include the hydraulic cements and mortars and sodium silicate cements. These have their own special techniques and are applied by their own skilled trades.

The *solvent-release adhesives* cure by evaporation of the solvent, thus allowing the consistency of the adhesive to increase. Many of the elastomeric cements belong to this type. The basic weakness of the solvent-release method is that the volume of the adhesive will be reduced by about one-third. This loss of volume sets up shrinkage stresses in the bond. These stresses may not be serious but at elevated temperatures lead to creep and weakening of the bond. Evaporation of the solvent also frequently results in voids in the layer of adhesive. The familiar white glues for household use are solvent-release polyvinyl acetate resins.

Resilient tile may use rubber-base mastics with a water or an alcohol solvent, as well as linoleum paste or an asphalt.

The *hot-melt adhesives* are a more recent development. These are mixtures of polymers applied hot to one surface; then the second surface is pressed onto the adhesive while it is hot. The hot adhesive wets both surfaces and cools to develop final strength. Although hot melts are new, the principle is as old as the hot-mopping of asphalt. Hot melts are used in the spiral-wound carboard tubes used in forming concrete (Sonotubes).

The *emulsion adhesives* consist of dispersed globules of adhesive in water. These bond by evaporation of the water.


## 17.3 chemically-curing

## thermosetting adhesives

The *chemically–curing thermosetting adhesives* used in construction and in construction materials include phenol–formaldehyde, urea–formaldehyde, melamine–formaldehyde, resorcinol–formaldehyde, and epoxy. Rigid foamed polyurethane is self-adhering, and some methods of installation use the layer of foam to bond the materials adjacent to the foam. Of all these materials, phenol–formaldehyde has the best weather and water resistance, which explains its use in exterior grades of plywood and curtain-wall construction. The other formaldehydes find interior applications in plywood and laminates.

The epoxies are more expensive adhesives but have made a reputation for themselves in applications where reliability, high strength, hardness, weathering, bonding capacity to ill-fitting surfaces, and resistance to water are required. They do not shrink on curing. Epoxies are sufficiently strong and hard to use as a filler for repairing holes in concrete floors, or for bonding new to old concrete. Such applications may require thick layers of adhesive: almost all other adhesives are required to be deposited in a thin film in order to develop maximum strength. The strength of an epoxy bond thus is independent of the thickness of the epoxy layer, and the material can perform a dual service as a void filler and an adhesive.

## 17.4   elastomeric adhesives

Most of the pressure-sensitive tapes such as Scotch tape use an *elastomeric adhesive*. Here we are concerned with the liquid elastomeric adhesives. These are usually solvent-release rubbers, the solvent being a gasoline or alcohol, or an emulsion in water. There are a great many varieties, and the following list does not include all types:

1. acrylic elastomers
2. acrylonitrile–butadiene (nitrile rubber)
3. styrene–butadiene (SBR rubber)
4. polyurethane

The nitrile rubber adhesives show good resistance to oils and solvents. They are used to bond PVC, insulating foams, and phenolic moldings.

Neoprene adhesives are *contact adhesives*; that is, they stick immediately on contact. They are well suited to the bonding of plastics, and decorative plastic laminates to wood surfaces. SBR adhesives are also contact types used in mastics and suited to the bonding of flooring materials to wood or concrete subfloors. However, SBR is not recommended for joints that must sustain tensile or shear forces of any magnitude.

## 17.5   application methods

In applying such coverings as floor tile, the adhesive or mastic is spread over the subsurface with a notched trowel. Sometimes the adhesive is brushed or rolled over the surface. The tile are laid in the wet mastic before it hardens. The notched trowel creates ridges in the mastic, and the tile must be laid before these ridges set up and so that the ridges are pressed out smoothly.

In contact bonding, both the surfaces to be bonded must be coated evenly with the contact cement. When the solvents have evaporated and the cement is dry to the touch, the two buttered surfaces are pressed together. A bond forms immediately on contact. Bonding pressure should be applied from the center to the sides, to remove trapped solvent.

## 17.6 characteristics of adhesives

A *substrate* is a material upon which a coating is spread. An *adhesive* is a material that will wet and join two surfaces by attachment to them. A *cement* is the same thing as an adhesive.

A catalyst is the second component in a two-component thermosetting resin or adhesive. It is a chemical that will react with the other component to set or cure it (note that this is not a chemist's definition of a catalyst). The curing is a cross-linking operation; the catalyst becomes incorporated into the adhesive and is not a solvent.

*Tack* is the property of an adhesive that causes a surface coated with the adhesive to form a bond on immediate contact with another surface. Tack is thus the stickiness of an adhesive.

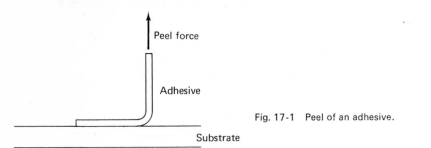

Fig. 17-1   Peel of an adhesive.

*Peel strength* means the resistance of the adhesive to stripping (Fig. 17.1) and is expressed in pounds per inch width. Although many adhesives have shear strengths exceeding 5000 psi, the peel strength of any adhesive is limited. Adhesive joints must always be protected against peel stresses. Adhesive joints also are relatively weak in tension. Most joint designs ensure that the adhesive is stressed in shear.

Test methods for adhesives are listed at the end of the chapter.

## 17.7 pressure-sensitive adhesives

*Pressure-sensitive adhesives* are usually made up in tape form on a backing material. Such an adhesive is a substance that holds after a brief application of light pressure, and its predominant characteristic is tack. The adhesive used is some type of elastomer compounded with tackifying substances so that the adhesive remains soft, viscous, and permanently tacky. Such adhesives should peel from the surface without leaving any residue.

Pressure-sensitive tapes do not wet the surface and therefore cannot be employed to carry any stress, even minor stresses. They have no resistance whatever to peel, and their useful life is limited to perhaps 1 year.

## 17.8 surface coatings

No decorative surface coating can be successfully applied to an improperly prepared surface. Therefore, the remarks of Section 17.1 apply with equal emphasis to surface coatings as to adhesives.

In Section 14.9 a comparison was made between the convertible or thermosetting finishes, paints, varnishes, enamels, and the nonconvertible or thermoplastic coatings, such as shellac and lacquers. The nonconvertible coatings are greatly inferior in durability and resistance to weather and are susceptible to solvent attack. Since they harden by loss of solvent, it is not possible to build a heavy coating of such materials when a third or more of the deposited film is lost by evaporation.

The convertible finishes harden by cross-linking. The cross-linking agent may be oxygen from the air, or a catalyst in the case of two-component systems, as with epoxy paints.

A *paint* is a mixture of a vehicle (binder) and a pigment or filler, A *varnish* is a vehicle with no pigment. An *enamel* is a blend of a paint and a varnish. A *lacquer* is a thermoplastic coating that hardens by solvent evaporation.

Paints were originally based on unsaturated oils such as linseed oil or tung oil. About 50 years ago oil-modified alkyd resins first began to replace the paints using natural oils only. Many present-day paint formulations contain no natural oils. More recently the emulsion or latex paints have predominated for interior applications, while vinyl acetate and acrylic latexes are increasingly used for exterior work.

An emulsion and an aerosol are similar in principle. An aerosol is a dispersion of liquid in a gas (usually the gas in aerosol cans is a Freon refrigerant). An emulsion is a suspension of fine particles of a liquid within another liquid. Latex paints are emulsions of organic materials in water. The spherical particles of paint are not soluble in water, but are simply distributed through the water medium. On application to the surface being painted, the water evaporates and the resin particles coalesce to form the paint film. Particle size in the emulsion is much too small to be seen with the naked eye.

The emulsion must remain stable with the passage of time, although the shelf life of emulsion paints is not indefinitely long. The size of the particles of paint tends to increase with age as small particles contact each other and combine. This coalescence of particles causes an increase in viscosity.

Most latex paints are based on styrene–butadiene.

## 17.9   paint formulation

A paint mixture may require the following ingredients:

1. vehicle or medium
2. pigment
3. extender
4. drier
5. dyes

The finely ground solids of the paint are technically known as the *pigment*, and the liquid portion is called the *medium* or *vehicle*. Pigments must necessarily be white so that the coloring agents (which actually pigment the paint) may be added to supply the actual color. *Driers* are added to accelerate formation of the film. *Extenders* increase the "body" of the paint and improve the abrasion resistance. *Dyes* are used when the surface coating must be transparent but colored. *Thinners* are volatile solvents used to adjust the volatility of the paint. Paint that is sprayed will require more thinner than paint that is to be brushed. *Volatility* roughly corresponds to the evaporation rate or drying rate of the paint. One of the disadvantages of a latex paint is the difficulty of adjusting its volatility. The thinner in such a paint is really the water medium, and water evaporates rather slowly compared with standard solvents.

Formerly white compounds of lead were the standard white hiding pigments for paints. Because lead compounds are toxic, these compounds are not now in favor. Titanium dioxide, which is now used, is far superior as a body or pigment, and is used in every kind of paint except dark-colored ones and paints for special purposes, such as aluminum paint. It is not toxic. Other pigment materials are zinc oxide, zinc sulfide, lithopone (a mixture of zinc sulfide and barium sulfate), and antimony oxide.

A good hiding pigment must first be white, together with a range of other requirements. For best whiteness, a high index of refraction is required, and titanium dioxide is superior to all others in this respect. The pigment must supply opaqueness or hiding power, so that any design on the substrate will not show through the paint. It must protect the paint resin against ultraviolet degredation; zinc oxide is the best pigment in this characteristic and is used in exterior paints. A number of other requirements of some technical complexity are also needed in the pigment. All these requirements are sufficiently numerous that blends of pigments are now employed in paints. The lead pigments are generally inferior in all respects to the others mentioned, but have characteristics useful for special paints: they form soaps which improve the wetting of undercoats and primer coats to steel.

Extenders control the gloss of the paint and adjust such characteristics as viscosity and brushability. Some of the common extenders are kaolin, barite (barium sulfate), diatomaceous silica, gypsum, and whiting (calcium carbonate). Whiting improves gloss, and diatomaceous silica reduces gloss.

### 17.10  paint resins

Alkyd resins are widely used in synthetic finishes. These materials are made by chemically reacting an alcohol and an organic acid, hence the name alkyd—alcohol–acid. Since linseed and similar oils are excellent vehicles for paints applied to wood, alkyds are combined with oils. These are called long, medium, and short oil resins, depending on the relative amount of oil used in the formulation:

up to 20 gal of oil per 100 lb of resin—short oil resin

20–30 gal of oil per 100 lb of resin—medium oil

30 or more gal—long oil

The long oil resins are slower drying, giving a softer film with better durability and flexibility, and are preferred for construction. The short oil blends are used for household appliances and furniture.

The *alkyds* are excellent materials for exterior paints, resisting ultraviolet radiation for years, and remaining flexible. Their adhesion and resistance to moisture are excellent, and they are frequently blended with other resins to improve these properties in paints.

*Acrylic-based paints* are also durable against ultraviolet radiation, and retain their gloss well. They are less commonly employed in construction. Both alkyds and acrylics are used in factory-baked enamel paints; acrylics are also used in latex paints.

The *epoxy paints* have all the virtues of epoxy adhesives: excellent adhesion, weatherability, hardness, abrasion resistance, and resistance to chemical attack. They make excellent masonry paints, since they are not attacked by the traces of alkali in concrete and brickwork.

*Urethane enamels*, like the epoxies, provide a superior surface coating, with excellent bonding, abrasion resistance, and resistance to moisture and chemical attack. They are excellent masonry paints. Other masonry paints are based on polyvinyl acetal resins.

The cellulose polymers, cellulose nitrate, cellulose acetate, and ethyl cellulose, are used in lacquers. They are not recommended for exterior applications, since they are thermoplastic vehicles. The other paint resins discussed here are cross-linked, but not the celluloses.

*Latex paints* use elastomer resins, chiefly SBR. These are popular because of the ease with which they may be cleaned up and their resistance to fading and peeling. They tend as they age to crack along the lines of the grain of wood. Latex paints are used for both exterior and interior use, and on masonry surfaces.

Clear and transparent finishes on exterior surfaces, especially wood, have a very short life. This is explained by the absence of pigmenting materials to absorb ultraviolet radiation.

*Fire-retardant paints* function by foaming into a heat-insulating mat that protects the underlying material from excessive temperature. Such paints do not wear or weather well and are not suited for exterior application.

## 17.11  paints for metal surfaces

While many paints adhere sufficiently well to steel surfaces, metal pose unique problems for paints. Unlike wood, concrete, masonry, and wallboard, metals have a nonporous surface. Steel has a chemically reactive surface, as is indicated by the readiness with which it rusts, and paint attaches to it. Adhesion to aluminum is more difficult, aluminum being a less reactive metal. Few paints will adhere to copper. In more severe environments the standard paints cannot give adequate protection to metals, and special coatings are formulated for this purpose. Bituminous coatings and coal-tar epoxies are necessary for buried steel tanks and pipes. Several heavy coats of epoxy paint give excellent resistance to chemical attack, but such treatment is not recommended for buried metal since backfilling may damage the paint film.

*Red lead* is a widely used anticorrosive primer for structural steel. The lead compound reacts with oils on the steel surface to produce lead soaps that are water-repellent. For aluminum, a thin coat of yellow zinc chromate is the usual protective primer.

Both zinc and aluminum metal are used in protective coatings for metals. Paints that incorporate zinc dust are used on galvanized surfaces and as a zinc-rich protective coating on steel.

*Aluminum paints* give excellent protection and adhesion to metals and are used for both priming and finishing coats. These paints have a dispersion of fine flakes of aluminum, smaller than 100 mesh, which forms an impervious

Fig. 17-2  Leafing effect of aluminum paint.

"leafing" film bonded together with the vehicle (Fig. 17.2). The vehicle is an asphalt, which provides good bonding. The combination of the moisture-proof asphalt together with the moisture-proof leafing effect makes this paint a very effective vapor barrier. It is also used as a vapor barrier on surfaces other than metals. It has superior hiding power: a film of aluminum paint 0.0005 in. thick will completely hide any surface.

*Vinyl coatings* are effective for protecting metals and are used on prefabricated steel buildings and aluminum siding. These are factory applications using hot-spraying to provide thicker films for longer life and better weather resistance.

### 17.12   paint testing

The *hiding power* of paints is tested by means of standard hiding-power charts, some of which are shown in Fig. 17.3. These sheets are printed with patterns in black and white. A film of controlled thickness in mils is applied to the chart. If the paint has the required hiding power, the pattern is obliterated.

Fig. 17-3  Selection of hiding-power charts. The one on the right has received a coat of paint for a hiding-power test.

There are several tests for flexibility and adhesion of surface films. The paint may be applied to a small panel of metal, and after drying of the paint the panel is bent 180 degrees around a bar of specified diameter. Alternatively a conical mandrel may be substituted, and the diameter of the cone at which cracking of the film first occurs is noted.

The impact test uses a steel ball with a 1-in. radius. This is dropped from varying heights and the inch-pounds of energy recorded at which the film separates. The direct impact test is made on the painted face of the steel panel; the reverse impact test is made on the unpainted side. The latter test is more severe.

Still another test is used for adhesion. A $\frac{5}{16}$-in.-diameter steel rod with a rounded end is used to mar the painted steel panel in two directions at right angles to each other. The rod is held perpendicular to the film and the rounded end drawn across the painted surface, using sufficient pressure to indent the film. The paint film is examined along both indentations, especially where they cross, for loss of adhesion.

Special apparatus is required for testing the abrasion, scratch, and weathering resistance of paint films.

## PERTINENT ASTM SPECIFICATIONS

D907-67 (Part 16) Terms Relating to Adhesives
D16-68 (Part 20) Terms Relating to Paint and Related Products
D1876-61T (Part 16) Peel Resistance of Adhesives (T-Peel Test)
D950-54 (Part 16) Impact Strength of Adhesive Bonds
D905-49 (Part 16) Strength of Adhesive Bonds in Shear
D903-49 (Part 16) Peel Strength of Adhesive Bonds
D897-68 (Part 16) Tensile Strength of Adhesive Bonds
D816-55 (Part 16) Testing Rubber Cements
D2197-68 (Part 21) Adhesion of Organic Coatings
D1310-67 (Part 20) Flash Point of Volatile Flammable Materials by the Tag Open-Cup Apparatus
D2651-67 (Part 16) Preparation of Metal Surfaces for Adhesive Bonding

## QUESTIONS

1  What is meant by wetting?

2  Why should a thermosetting surface finish be superior to a thermoplastic type?

3  Why must masonry and concrete surfaces be fully cured before application of paint?

4  Why are oil paints unsuited to concrete surfaces?

5  What is the chief weakness of glues as adhesives?

6  What disadvantages are characteristic of a solvent-release type of adhesive?

7  List the outstanding characteristics of epoxy adhesives.

8  What is a contact adhesive?

9  What is a pressure-sensitive adhesive?

10  Define tack.

11  Define peel strength.

12  Differentiate between a convertible and a nonconvertible surface film.

13  Explain the difference between a paint and a varnish.

**14** What is an emulsion paint?

**15** Explain the function of the vehicle and the pigment in a paint.

**16** Define volatility.

**17** Why would a lead-base paint not be allowed in painting the walls of a cold storage meat freezer?

**18** Why is titanium dioxide favored as a pigment in paints?

**19** Why is zinc oxide used as the pigment in exterior paints?

**20** Explain why lead oxide is used as a pigment in priming coats for structural steel.

**21** What is the meaning of a "long oil" resin?

**22** What materials are used for pigment and vehicle in an aluminum paint?

**23** What is meant by the term "hiding power"?

**24** Which will increase gloss in a paint: increased vehicle or increased pigment?

**APPLICATIONS**

*Part* **3**

The roof is the most complex problem in the application of construction materials to a building, and most types of construction materials are used in roof structures. The roof is exposed to the widest range of environmental hazards. So as best to display the application of materials to the problem of a roof, this chapter will discuss first the built-up roof, then more recently developed types of roofing membranes, and finally other standard types of roofing.

## 18.1    the roofing problem

The function of a roof is to seal the building against the elements and to maintain this seal over a considerable number of years. The successful execution of this function is surprisingly difficult. The roof must resist ultraviolet radiation, oxidation, weather conditions such as rain, wind, and abrasion, wind pressure, snow load, hail storms, high temperatures, low temperatures, the temperature differential across the roofing membrane from the inside to the outside, the daily temperature variation, moisture transmission and con-

# ROOFING

# 18

densation of moisture within the roofing membrane, and the tendency of roofing materials to delaminate.

Long as it is, this list of roofing hazards is incomplete. The roof design can at best be no more successful than the building design. A poor foundation may be a cause of building movement which the roof cannot successfully accommodate. The building must expand and contract with temperature changes, and the roof must accommodate these changes. They are, of course, more severe in larger buildings. All these problems are complicated by the occasional deficient architectural specification, the desire to economize on materials for the roof, and the occasional faulty workmanship. It is not surprising, therefore, that everyone knows of a leaky roof. Such roof failures are more common in severely cold regions; in the Canadian north roofs have failed after two winters, chiefly due to heavy deposit of ice within the roof membrane.

Roofing is a major industry. About 80 square miles of built-up roofing is installed annually in the United States. This volume of business, plus the roofing problems involved, has generated some aggressive development of new roofing materials.

## 18.2  the elements of a roof

A roof requires three basic elements. First a *deck* is required to support the roof membrane. This may be concrete, wood, steel sheet, or other materials. Second, a *thermal barrier* of insulating material is usually applied to the deck. To this may be added a vapor barrier to protect the insulating and roofing materials from moisture condensation. Finally, the impervious *roofing membrane* seals the roof structure.

Each of these materials may have an unfavorable effect on the performance of the others. The use of insulation underneath the roofing membrane will make the roof hotter in summer and colder in winter, since it will prevent exchange of heat from the interior of the building to the roof and vice versa. Roof surface temperatures of 150°F on insulated roofs are not unusual in the summer months. The vapor barrier may trap moisture in the roofing materials. For example, a poured concrete roof deck requires a long time to dry out. The general contractor or the owner may want the roofing installed before the deck has lost its moisture, and the result may be a crop of blisters in the roof membrane caused by this moisture (if moisture is trapped and evaporates within the roof, it will expand about 1000 times).

The roof therefore is a remarkably complex sandwich system. The flashings of the roof are not a component of this sandwich system but are a requirement for sealing joints at the edges of the roof, the openings cut into the roof, expansion joints, and other places.

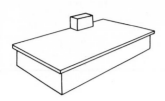

Flat Roof

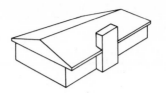

Gable Roof

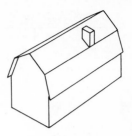

Gambrel Roof

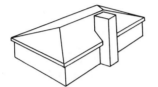

Hip Roof
(Cottage Roof)

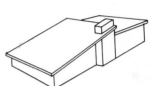

Shed Roof

Fig. 18-1   Roof shapes.

## 18.3   roof insulations

The usual roof insulation boards are the following:

1. Compressed and baked cork $\frac{1}{2}$–6 in. thick; $K$ factor 0.26.
2. Cellular glass, closed cells, with kraft paper; $K = 0.39$, and perm rating zero. No vapor barrier therefore is required with this material. Thickness is $1\frac{1}{2}$ or 2 in.
3. Fiberglass with asphalt-laminated kraft paper, thickness from $\frac{7}{16}$ to $2\frac{7}{16}$ in; $K = 0.25$.
4. Perlite with mineral fiber and binder, top surface coated with asphalt. Thicknesses $\frac{3}{4}$–3 in., $K = 0.36$.
5. Fiberboard using wood or cane fiber, plain or impregnated with asphalt, coated with asphalt or uncoated. Thicknesses $\frac{1}{2}$–3 in.; $K = 0.36$.
6. Styrofoam board, thicknesses $\frac{1}{2}$–3 in., coated or uncoated; $K = 0.24$.
7. Rigid foamed polyurethane board, thicknesses $\frac{1}{2}$–3 in., $K = 0.16$.

Both the styrofoam and the urethane board are sometimes sandwiched between faces of coated roofing felts.

The $K$ factors given are in the usual units of Btuh/ft³/°F/in. of thickness.

Wood and vegetable types of insulation are more vulnerable to moisture, one of the worst enemies of roofs, and may in time rot due to the continued

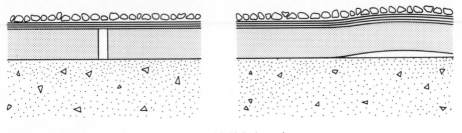

Fig. 18-2 Thermal strain in roof insulations with high thermal expansion may produce gapping in cold weather and bowing in hot weather.

presence of moisture. The very light styrofoam and polyurethane insulations, 2 pcf, are easily crushed by the freezing of moisture in the insulation.

The roof insulation may have four functions:

1. To provide a suitable base for the roof membrane.
2. To flex in order to accommodate the difference in strain, thermal and other, between roof membrane and roof deck.
3. To provide a fire-resistant separation between deck and membrane.
4. To insulate the building. Steel decks especially require insulation before the roof membrane is applied.

The insulation does not experience the wide temperature ranges of the roofing membrane. Nevertheless, its thermal expansion may be a significant property in determining the success of a roof. Consider styrofoam and polyurethane board insulation, which have a thermal expansion in the range of 0.00005 in./in./°F, and suppose that such board in not bonded to a membrane or deck. A temperature change of 100°F will result in a change of length in an 8-ft insulating board of about $\frac{1}{2}$ in. The board must either bow or shrink, as shown in Fig. 18.2. If the joints between insulating boards open, the roofing membrane can crack. Such thermal strains are not so severe in practice because both the roofing membrane and the roof deck will also expand and the bond between these elements will restrain to some degree any extreme movements. On the other hand, such movements may break the bond between roof elements, or the bonding technique may be inadequate. Despite its limitations in other respects, fiberboard has a very low thermal expansion and does not create such problems in thermal strain.

Fig. 18-3 Concrete roof plank.

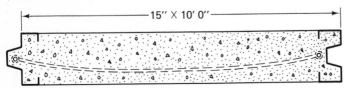

15″ X 10′ 0″

Water-vapor migration causes no problems in a roof until condensation occurs. The vapor barrier is supplied to prevent such trouble. On the other hand, the vapor barrier may trap moisture within a roof system, thus generating a new set of problems. It is best to omit the vapor barrier if this can reasonably be done, as it can in warmer climates. It is very important to bear in mind that vapor barriers are not true barriers, only vapor resistances. A true vapor barrier has a perm rating of 0.00; all others admit some water vapor, especially if there are joints or punctures. Calculations of the amount of water vapor that passes through most vapor barriers in an extended period of time such as a year lead to somewhat sobering figures.

Roofing asphalt is a good vapor barrier and is adhesive. Laminated kraft paper, made of two layers of heavy kraft paper, is often used. Its perm rating is 0.25. Some vapor-barrier systems used in walls cannot be applied to roofs. Polyethylene film is applied to wall studs, but the use of this film under polyurethane foam on a roof is not permissible, since the foam will not adhere to polyethylene.

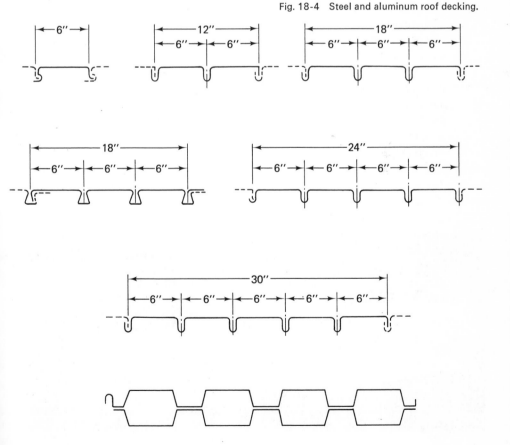

Fig. 18-4 Steel and aluminum roof decking.

Styrofoam as a roof insulation has some limitations. Hot asphalt can melt styrofoam, and the solvents in asphalt will dissolve the foam.

Some method must be used to attach roofing materials to the roof decking. Roof decks may be nailable or nonnailable. Wood, wood fiber, and gypsum decks are nailable; concrete and metal decks are not. Roof materials are usually anchored to nonnailable decks by bitumens, either hot-mopped or cold applied, although special mechanical fasteners may also be used.

Typical concrete roof planks are shown in Fig. 18.3. Various types of steel and aluminum roof decking are given in Fig. 18.4. Concrete plank has plain or tongue-and-groove edges. Metal decking is joined by lapping adjacent pieces of decking.

## 18.4  roof forms

The common roof styles for residences are shown in Fig. 18.1. Commercial and industrial buildings may have flat, pitched, or contoured roofs.

The *flat roof* is subject to the ponding of water (Fig. 18.5). Should the roof membrane crack, water can be admitted into the roof and building. Flat roofs are also subject to traffic and often must carry ventilating equipment. This type of roof is the most difficult to design and specify.

Ponding of water on a flat roof has sometimes led to collapse of the roof. If water ponds on the roof, the roof deck deflects thus increasing the depth of the pond. This leads to still further deflection and increasing depth of water, until the roof collapses.

The *pitched roof* can shed water, but ice tends to build up at the eaves, which receive little heat from the interior of the building. Such ice accu-

Fig. 18-5  Built-up roof. Ponding occurs toward the right-hand side of the roof.

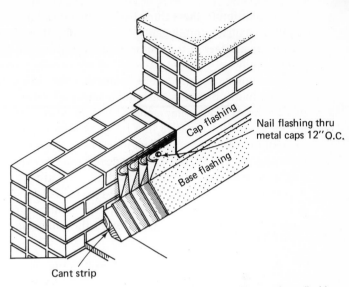

Cant strip

Fig. 18-6 Base and cap flashings.

mulation will dam melting water on the roof, though on a more limited scale. Just as a pitched roof can shed water, so can its asphalt sag and flow if the asphalt is not of proper consistency.

*Contoured roofs* have become more common. These may be of many shapes, circular, paraboloid, hyperbolic, folded plate, geodesic dome, and many others. The use of newer plastic and rubber roofing materials is often advantageous on such roofs, which are less well suited to standard roofing methods.

## 18.5   the built-up roof

A *built-up roof* membrane has three components:

1. The bitumen, which provides the impermeable seal against the elements.
2. The roofing felts, which reinforce the bitumen and prevent it from flowing.
3. The aggregate surfacing, normally a gravel or crushed rock.

This mineral surface protects the bitumen from the damaging effects of ultraviolet radiation and oxidation. It protects also against foot traffic, wind abrasion, and flame spread. Smooth-surfaced asphalt roofs, which usually incorporate asbestos or fiberglass felts, may omit the aggregate surfacing; such a roof does not weather quite so well. All aggregate should be clean and dry and sized from $\frac{1}{4}$ to $\frac{3}{4}$ in. A minimum of 400 lb of gravel or 300 lb of blast furnace slag is required per square.

In addition to these three basic elements, a built-up roof, like any roof, requires auxiliary components called flashing (Fig. 18.6). *Base flashings* are the upturned edges of the roof membrane at its edges. *Cap flashings* cover the exposed joints of the base flashing and are usually made of such sheet metals as aluminum, galvanized steel, stainless steel, copper, or lead.

The built-up roof is basically a sandwich series of two to five (usually two or three) asphalt-impregnated felts bonded with layers of asphalt. The mopping asphalts are used in a range of softening points and viscosities for the requirements of roofs of various slopes. An asphalt with a low range of softening point, 135–150°F, is a "dead level" or type I asphalt. For a steep roof a "steep" asphalt must be used. Since coal-tar pitch cold-flows readily, it is limited to roof slopes of $\frac{1}{2}$ in./ft or less. This material has excellent self-healing for closing cracks, however. The significant properties of the several types of roofing bitumens are given in the following table:

| | Type I: Dead Level | Type II | Type III: Steep | Type IV: Special Steep | Coal-Tar Pitch |
|---|---|---|---|---|---|
| roof slope (in./ft) | $0-\frac{1}{2}$ | $\frac{1}{2}-3$ | $\frac{1}{2}-6$ | $\frac{1}{2}$–steep | $0-\frac{1}{2}$ |
| softening point (°F) | 135–150 | 160–175 | 180–200 | 205–225 | 140–155 |
| penetration, 100 g, 5 sec, 77°F | 18–60 | 18–40 | 15–35 | 12–25 | |
| ductility, 77°F | 10 min | 3 min | 3 min | 1.5 min | |

*Organic felts* are made from felted paper, shredded wood fiber, and similar materials, and are saturated with coal-tar pitch or asphalt. These are the cheapest and most popular roofing felts. Asbestos felts are the best type for resisting the various types of deterioration to which roofing felts are subject and are fire resistant. Fiberglass is also used.

*Saturated felts* are those that are saturated with bitumen. Coated felts are saturated and also coated with a film of asphalt. Coated felts are chiefly used as base felts. Mineral-surfaced felts also have mineral aggregate on the exposed surface. All these felts are 36 in. wide.

The organic felts are described by numbers that give the approximate weight of the felt per paper measure of 480 ft². This weight is the weight before being saturated with bitumen. The weights are 15, 30, 35, and so on. The inorganic felts are described in pounds per square.

*Cold-process bitumens* are cutback or emulsion asphalts applied without heating. These are popular for repair and maintenance of roofs because heating kettles are inconvenient to set up for small jobs. Cold-process materials are also used in new construction under special conditions, such as air pollution regulations that may forbid kettles. Such bitumens are more expensive than hot-mopped materials.

In summary, a built-up roof begins with a coated base sheet, which is also a vapor barrier, mopped to the deck or insulation. Hot-mopped steep asphalt is used for the base sheet. On top of the base sheet are applied one or more plies of saturated roofing felt which are shingled as shown in Fig. 18.7, and hot-mopped. The top coat of bitumen is the flood coat, in which is embedded the surfacing aggregate.

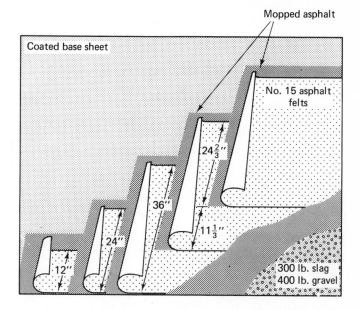

Fig. 18-7   Three-ply built-up roof with No. 15 felts.

Techniques are available for spraying asphalt roof coverings. A special gun with three nozzles and a fiber chopper is employed. The gun chops fiberglass, which is sprayed from the center nozzle. The two side nozzles deliver an asphalt emulsion which coats the fiber and transports it to the roof surface to produce a film of reinforced asphalt. The method is useful for roofs of irregular shape, for minor repairs and recoating, but can also be used with standard roofing felts.

## 18.6   newer developments in roofing membranes

Recent years have seen the increasing use of plastics and synthetic rubbers applied either in fluid or sheet form to produce roofing membranes. These new materials reduce the weight of the roof and reduce the labor cost. Roof membranes weighing as little as a tenth of a pound per square foot are possible with these new materials. They have excellent elongation, sometimes as high as 450 per cent. They can be applied to any shape of roof, however complex or curved, or even to a vertical surface, and are usually easy to repair. They are often sprayed over existing built-up roofs. They cannot, however, be used for bridging over cracks.

The most highly developed of these new membranes is the sprayed roof of polyurethane foam coated with butyl rubber. At least 1 in. of polyurethane must be sprayed, using a 2-lb density, although a 4-lb insulation has greater strength, durability, and resistance to ice formation in the foam. An Elastron, Irathane, or other proprietary fluid rubber priming coat must be sprayed on the deck if a moisture barrier is required. On steel roof decks, certain proprietary priming coatings are available to prevent any oil film on the steel from interfering with the adhesion of the foam to the deck. Such adhesive is also required if the urethane is sprayed against a wood deck, since the foam

Fig. 18-8 A sheet of 1-mil polyvinyl fluoride (Tedlar) covers the sheet of paper over the "Tedlar" heading. Part of the PVF film has been painted with white Hypalon roofing rubber.

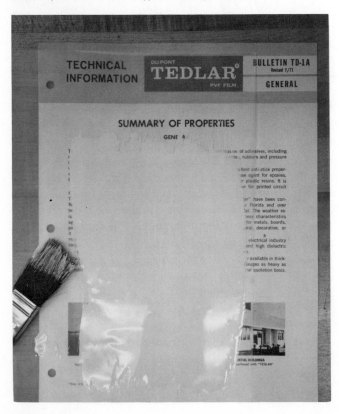

could be loosened if the wood were to take up moisture in a significant amount. Otherwise the urethane foam gives excellent adhesion to materials.

The surface of the foam must be protected from ultraviolet radiation. The protective coatings serve this function and also protect the soft foam from abrasion or foot traffic. In Chapter 14 it was pointed out that most elastomers are unsaturated and therefore subject to attack from ozone in the air. These elastomeric roof coatings must be saturated or otherwise compounded for ozone and oxygen resistance. Usually either Hypalon (chlorosulfonated polyethylene) or butyl rubbers such as Elastron are used. Such coatings are applied as viscous fluids thinned with solvents. After the solvent evaporates the elastomer will cure to a tough abrasion-resistant and fire-resistant film. A dry film thickness of at least 15 mils is required, which corresponds roughly to 30 mils wet. Two coats are required, since the evaporation of the solvent may produce occasional pinholes in the elastomeric film.

Elastron butyl rubber is available only in black, gray, or tan. Hypalon can be given any color, including white and pastel shades (Fig. 18.8), and produces a beautiful roof with a gloss finish. It cannot be applied more than 4 mils thick in a single coat, and therefore is usually applied as a topcoat to a cheaper and more convenient elastomer base, such as neoprene (chloroprene). Neoprene itself is not as durable as Hypalon as a topcoat, but the Hypalon provides both beauty and weatherability to the roof.

Polysulfide (Thiokol) rubber has been used for the same roofing applications as Hypalon, and is available in a range of colors. It can be deposited in heavier thicknesses than Hypalon. Silicone rubber also has excellent properties as a roofing material, but is expensive. It would be superior to the other elastomeric membranes in cold climates, since these lose elasticity at low temperatures. At the time of writing, acrylic elastomers are being introduced as roof coatings, under trade names such as Diathon.

The sheet membranes are more expensive and more difficult to install successfully. A sheet membrane, however, is better adapted to the bridging of cracks. Sheet neoprene was used for the roof of Dulles International Airport at Washington, D.C., using two coats of liquid Hypalon for top coating. Butyl rubber, Hypalon, and neoprene–hypalon are also in use as sheet membranes for roofs.

Polyvinyl fluoride (Tedlar) film has been used as a top surface for roofs. It has remarkable weatherability and abrasion resistance and protects the material underneath from ultraviolet degredation.

### 18.7 roll roofing

Roll roofing is roofing material put up in rolls (Fig. 18.9). The material is a heavy asphalt-impregnated felt paper in 36-in. widths. Four types are available:

1. *Smooth:* covered with a smooth asphalt coat.
2. *Mineral surface:* This type is heavier than the smooth roll, with a layer of crushed mineral in the surface.
3. *Pattern edge:* This is mineral-surfaced except for a band down the center. The roll is partly cut along this strip so that it will separate into two 18-in.-wide patterned strips. These are lapped 2 in. when installed so that only the mineral surface is exposed.
4. *Selvage roll:* This is mineral covered for a width of 17 in. and is lapped 19 in. on installation, so that only the mineral surface is exposed.

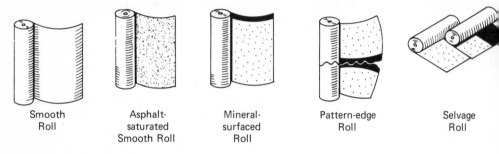

Smooth Roll     Asphalt-saturated Smooth Roll     Mineral-surfaced Roll     Pattern-edge Roll     Selvage Roll

Fig. 18-9   Roll roofing.

The back side of roll roofing is dusted with a mineral powder to prevent the coil from sticking together.

## 18.8   shingles

Shingles require a minimum slope to the roof of at least 3 in 12 if they are to shed water.

*Asphalt shingles* are produced from heavy organic felts saturated and coated with asphalt. Mineral granules are pressed into the coating on the exposed surface. Asphalt shingles are produced in various weights, styles, and colors, the weights being given in pounds per square of roof coverage. Galvanized large-head roofing nails are used to fasten them. Such shingles are not flexible and can be lifted and broken by high winds. Figure 18.10 shows several patterns of asphalt shingle, one type carrying an adhesive to prevent lifting by winds.

*Asbestos–cement shingles* are made of asbestos fiber in portland cement paste, surfaced with mineral granules in a range of colors. They are also available in textured surfaces. This type of shingle is very hard, and nail holes are predrilled. Cutting of asbestos–cement shingles must be done with a special shear.

*Aluminum shingles* are perhaps less commonly seen. Both plain, anodized, and colored enamel finishes are offered.

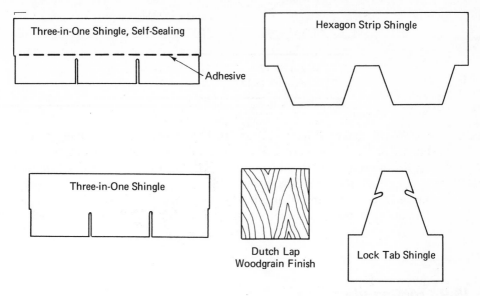

Three-in-One Shingle, Self-Sealing

Adhesive

Hexagon Strip Shingle

Three-in-One Shingle

Dutch Lap
Woodgrain Finish

Lock Tab Shingle

Fig. 18-10   Asphalt shingles.

*Wood shingles* are made of several types of softwood, but western red cedar is preferred because of its light weight, durability, and straight grain and relative absence of knots. This wood also does not have the severe expansion and contraction due to moisture changes characteristic of most woods, and hence is less likely to split and cause a roof leak.

Wood shingles have a tapered shape because they are quarter-sawed (Fig. 18.11). The widths of the individual shingles are random, but no shingle may be wider than 14 in. Lengths are 16, 18, and 24 in. Grades are Nos. 1, 2, and 3, No. 1 being the superior grade.

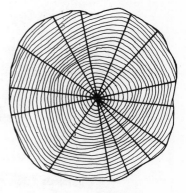

Fig. 18-11   Quarter-sawing of a log for shingles.

The exposed length of a wood shingle depends on both the slope of the roof and the length of the shingle:

| Pitch of Roof | Length of Shingle (in.) | | |
|---|---|---|---|
| | 16 | 18 | 24 |
| $1\frac{1}{2}$ or 2 in. in 12 | $3\frac{3}{4}$ | $4\frac{1}{4}$ | $5\frac{3}{4}$ |
| $2\frac{1}{2}$ in. or more in 12 | 5 | $5\frac{1}{2}$ | $7\frac{1}{2}$ |

Wood shakes (Fig. 18.12) are perhaps more attractive than shingles, but are also more expensive. Shakes are split from cedar blocks rather than sawed, to produce a rougher, more irregular, and more interesting face than a shingle shows. They are either tapered or straight. Lengths for tapered shakes are 18, 24, and 32 in., in thicknesses of $\frac{1}{2}$ to $\frac{3}{4}$ in. and $\frac{3}{4}$ to $1\frac{1}{4}$ in. Straight-split shakes are made in 18- and 24-in. lengths and $\frac{3}{8}$ in. thick.

## 18.9   roofing tile

*Roofing tile* is another terra-cotta product designed for roofing applications. Such tile are heavy, about 1000 lb per square, and can be supported only on

Fig. 18-12   Shakes on the front of a professional office.

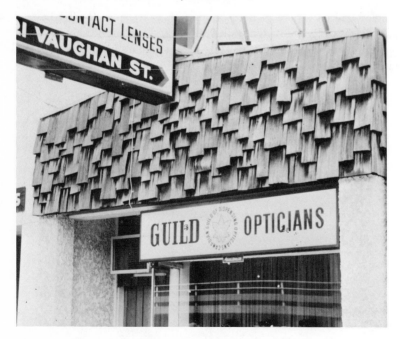

a strong roof deck. Roofing tile are made in a variety of attractive styles, some of which are illustrated in section in Fig. 18.13. Both glazed and un-glazed surfaces are available. Such tile are laid over an asphalt base, fastened with nails, and caulked. A variant of the terra-cotta tile is the concrete roof tile, which is a flat tile.

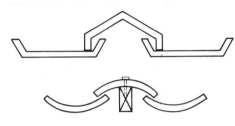

Fig. 18-13   Types of clay roofing tile.

Slate rock is easy to split into thin slabs suitable for roofing slabs, and is still used in those parts of the world where slate is available. Slate tile must be predrilled for nail holes.

## 18.10    sheet-metal roofing

*Sheet-metal roofing* is the usual roof for prefabricated metal buildings but is not confined to these structures. Metal roofing sheets are made in various gages, and if laid over supporting purlins, longer spans require heavier gages. Joints may be prefabricated. Alternatively the metal sheet is corrugated and a sufficient joint is produced simply by an overlapping of sheets, if the roof is sloped.

Sheet-steel roofing is galvanized. The steel sheet gages are given in a table in Section 11.12. Widths are 24, 30, and 36 in. Virtually any length can be provided that can be shipped, since these sheets, whether flat or corru-gated, are produced in continuous lengths in manufacture. Some of the cross sections of corrugated sheet are shown in Fig. 18.14, with gage and span requirements.

Next to steel, aluminum is the preferred metal for roofing. Although somewhat more expensive and rather soft in the roofing alloys, its appearance, workability, and corrosion resistance are superior to steel. The thermal expansion of aluminum, however, is twice that of steel. Best corrosion resistance is given by the unalloyed metal or by alloy 3003 containing man-ganese, and these are therefore used for roofing and flashing. Gage numbers are very slightly different from those for steel: 16-gage steel measures 0.0598,

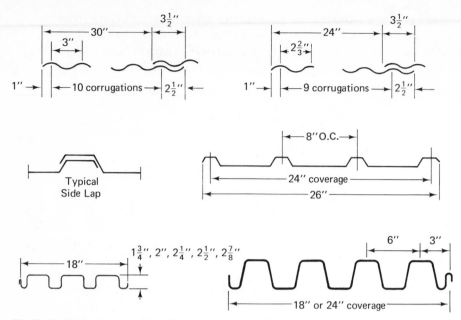

Fig. 18-14  Typical sheet metal roofing panels in section. These
are usually available in 26-24-23; and 22-gage thicknesses.

Fig. 18-15  Steel roof cladding with a plastic coating as applied
to the house shown in Fig. 19-19.

Fig. 18-16  Copper roofing on a church.

while 16-gage aluminum measures 0.0625. Aluminum can be welded and is solderable with special aluminum solders. These operations are rarely performed on the jobsite, however.

Care must be taken that aluminum does not contact other metals on a roof. Galvanic corrosion will result if different alloys of aluminum are in contact with each other and with moisture. Rivets and nails for aluminum must therefore be made of the same alloy as the sheet.

Copper is a traditional roofing material (see Fig. 18.16). It is readily solderable in the shop or even on the job but is a heavy metal. It weathers to a dark brown or a blue-green patina, although surface treatments are reasonably successful in maintaining its original color. In long lengths provision must be made for its thermal expansion, which is 50 per cent greater than that of steel or concrete.

Lead is rarely used as a roofing sheet. It costs about 35 cents per pound, and 1 ft$^2$ of sheet lead $\frac{1}{16}$ in. thick weighs 1 lb. Lead is highly corrosion resistant, except in contact with concrete, from which it should be separated by a film of asphalt.

Sheet zinc is occasionally used for flashings.

Corrugated galbestos sheet is a laminated sheet. Its core is a sheet steel. This is coated with molten zinc with a layer of asbestos felt pressed against the zinc while molten. The asbestos felt is saturated with asphalt. A colored waterproof coating is finally applied to both sides. Joints are made by side and end laps. Galbestos sheet is 12 ft long.

*Asbestos–cement sheets* are corrugated for roofing and decking. Decking sheets are used to support built-up roofs. Roofing sheets are applied either over a steel frame or a solid deck. Side joints are lapped one corrugation, end laps are 6 in. The sheets are fastened with nails, screws, or special types of fasteners, using predrilled holes.

*Corrugated plastic sheet*, either fiberglass-reinforced polyester or polyvinyl chloride without reinforcing, may replace other types of roofing sheets to admit light to a building at intervals along a roof. Both types of sheet are translucent. If desired, a whole roof may be covered with such sheets. The transparent plastics such as acrylics and polycarbonate are also used in skylights. Corrugated glass is also occasionally employed as a roofing material.

### PERTINENT ASTM SPECIFICATIONS

C553-64 (Part 14) Mineral Fiber Blanket and Felt Insulation

C578-65T (Part 14) Preformed Cellular Polystyrene Thermal Insulation

C591-66T (Part 14) Rigid Preformed Cellular Urethane Thermal Insulation

E108-58 (Part 14) Fire Tests of Roof Coverings

D1866-64 (Part 10) Translucency of Mineral Aggregate for Use on Built-Up Roofs

D2523-66T (Part 11) Testing Load-Strain Properties of Roof Membranes

D312-64 (Part 11) Asphalt for Use in Constructing Built-Up Roof Coverings

### SPECIFICATION WRITING

Write your own specifications for a first-quality roof, without regard to initial cost, using the following conditions and the precautionary remarks of this chapter. Consult manufacturers' literature and some sample roof specifications. You may need to set up additional requirements to define circumstances more closely.

The building has a roof area of 8000 ft². The roof is pitched $2\frac{1}{2}$ in. in 12. The location is eastern Montana, which can be very cold and windy.

1. Write a specification for a built-up roof with some type of board insulation. Decide your own perm requirement for vapor barrier.
2. Write a specification for a sprayed urethane insulation and elastomer roofing membrane. Decide your own perm requirements for vapor barrier. Protect against the following installation faults: spraying a roof deck that is wet with dew, spraying in the rain, spraying when it is too cold, spraying an irregular thickness or an uneven surface, leaving the

urethane exposed to sunlight for several days, and damage to the urethane foam from cleated or hard shoes.

## QUESTIONS

1  Obtain samples of Hypalon rubber in the can, Elastron rubber, and polyvinyl fluoride (Tedlar). Apply and test these materials. Apply a coat of Hypalon with a brush and measure the dry film thickness deposited. Apply a second coat and measure the double film thickness.

2  Why would heavy roofing felt be laid under asbestos–cement shingles and terra-cotta roofing tile?

3  Why is it preferable to avoid a flat roof?

4  Survey your geographical area for leaky roofs and obtain case histories. What conclusions can you draw?

5  Insulation imposes more severe service conditions on a roof. Explain.

6  A vapor barrier may create problems as well as solve problems. Explain.

7  List roof insulations in their order of decreasing thermal efficiency.

8  What are the several functions served by a roof insulation?

9  What functions are served by the bitumen, the felt, and the aggregate in a built-up roof?

10  Differentiate between a base flashing and a cap flashing.

11  What is a steep asphalt?

12  (a) Why must a polyurethane foam roof be coated with an elastomer? (b) What requirements are imposed on such an elastomer? (c) What elastomers have been employed for this purpose?

13  What advantages does aluminum have over steel as a roofing sheet?

14  A supplier cannot fill your order for a certain alloy of aluminum sheet and makes up the full order by mixing two aluminum alloys. Will you accept or reject the shipment? If you reject, on what grounds do you reject?

15  An asphalt built-up roof is a vapor barrier. Discuss the merits of an "upside-down" roof. In this type of construction, the roof insulation is placed on top of the built-up membrane. A layer of crushed stone protects the insulation from ultraviolet degredation.

### 19.1  wall as a building element

The traditional function of an exterior wall is that of the roof: to protect the interior of the building against the elements. A more modern point of view would state this differently: the function of an exterior wall is to act as a boundary to permit the maintenance of a controlled environment within the space enclosed by the building. A curtain wall of glass, for example, does not protect the interior of the building against heat and sunlight, or against darkness. Again, insulation in an exterior wall does not protect against the elements: the heating system does that. Insulation in an exterior wall has the function of reducing heating costs.

The exterior wall has other functions. Some exterior walls must be load-bearing, supporting some of the weight of the building. Exterior walls of multistory buildings, however, are non-load-bearing curtain walls, acting simply as cladding for the building. Finally, the exterior walls have an esthetic function. They should present a pleasing appearance, and in addition they sometimes identify the inhabitant by a distinctive design.

All walls, exterior and interior, may serve an even wider range of functions:

# EXTERIOR
# AND INTERIOR
# WALLS

# 19

1. *Partition walls:* interior walls that divide the interior into rooms or individual spaces.
2. *Retaining walls*, garden walls, and fences: serve purposes related to the property on which the building is set.
3. *Firewalls:* intended to restrict the spread of fire within a building. They must be of approved materials and installation and must be continuous over the building cross section, beginning at the foundation and extending above the roof.
4. *Party walls:* separate two adjoining buildings and carry the floors of each building on the single party wall.
5. *Parapet walls:* walls that extend above the roof of the building.
6. *Foundation walls:* built below grade to support the first floor of the building. They must be designed to withstand both the vertical loads of the weight of the building and also side thrust from backfill.
7. *Load-bearing walls:* must support beams, joists, floor loads, or roof loads.
8. *Curtain walls* or panel walls: carry none of the weight of the building except their own weight. A curtain wall is supported on the structural skeleton of the building. Its function is similar to that of a roof, and it must weather well. It can be made of a remarkable range of materials. Like the roof membrane, it must accommodate movements due to thermal expansion. Since the obvious module for a curtain wall is the story height, it must contain joints at every story of the building.

Four basic methods are used in the construction of walls:

1. a. *Solid walls:* consist of a solid mass of one type of construction material, such as stone, brick, or concrete.
   b. *Hollow walls:* constructed with materials containing hollow spaces, such as hollow tile walls. If the interior of the wall is a continuous open space as in Fig. 19.1, the wall is called a *cavity wall.*

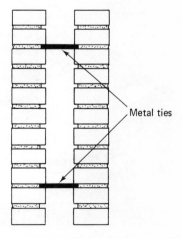

Metal ties

Fig. 19-1 Cavity wall reinforced with brick ties.

2. *Veneer walls:* use an outside facing surface of masonry or stone upon the main wall material. The veneer does not carry any wall loads.
3. *Curtain walls:* usually prefabricated modular units of wall.
4. *Stud walls:* framed with wood studs and stiffened with sheathing.

In this chapter the several types of wall construction will be discussed under four general headings: (1) light wood framing, (2) masonry bearing walls, (3) curtain walls, and (4) interior finishing materials.

## LIGHT WOOD FRAMING

### 19.2  wood frame wall

*Wood stud walls* are usually framed with 2 × 4s. Steel studding may be used in partition walls. Building codes require that exterior wall frames be stiffened against racking either by a 1 × 4 diagonal at each end of the wall or by

Fig. 19-2  Example of new trends in construction materials:
Steel roof trussed for house construction. The wall studs are also
steel. A general view of the house is given in Fig. 19-19.

plywood or other suitable wide sheathing. If the sheathing is board lumber, it is applied diagonally for racking strength and the 1 × 4s are omitted. If fiberboard or gypsum board is used for sheathing, building codes may required diagonal bracing. Fiberboard and gypsum cannot be used as a nailing base for exterior finish; for this purpose nailing strips or other special methods are required. Sheathing is the preferred method for racking strength in new construction. The sheathing also provides a resistance to heat loss and may serve as a base for exterior finish.

Wood framing uses either southern pine, spruce, or hemlock; fir is also used but is harder to nail. The western or platform type of building frame is preferred nowadays. In this construction each storey of the building is erected on its own subfloor, that is, all studs are of one-story height. The horizontal wood sills that support the walls are anchored to the foundation with anchor bolts every 8 ft maximum, as shown in Fig. 19.3.

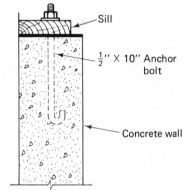

Fig. 19-3 Anchoring of sills to foundation.

A layer of building paper is applied between the sheathing and the exterior finish. The vapor barrier, if any, must be applied on the interior surface of the studs, that is, on the high-temperature side of the wall. Construction of a typical wall section for platform construction is shown in Fig. 19.4 and typical framing details at the first floor and second floor in Fig. 19.5.

### 19.3 stucco

*Stucco* is an exterior type of plaster made with portland cement. It will bond to masonry walls, but not to wood sheathing or gypsum board. To ensure attachment to a wood wall, either wire mesh or expanded metal must be used to key the stucco in place. The openings in the mesh must be sufficiently large to allow the first coat to penetrate to the wood base and encase the wire mesh. The mesh must not lie close against the sheathing.

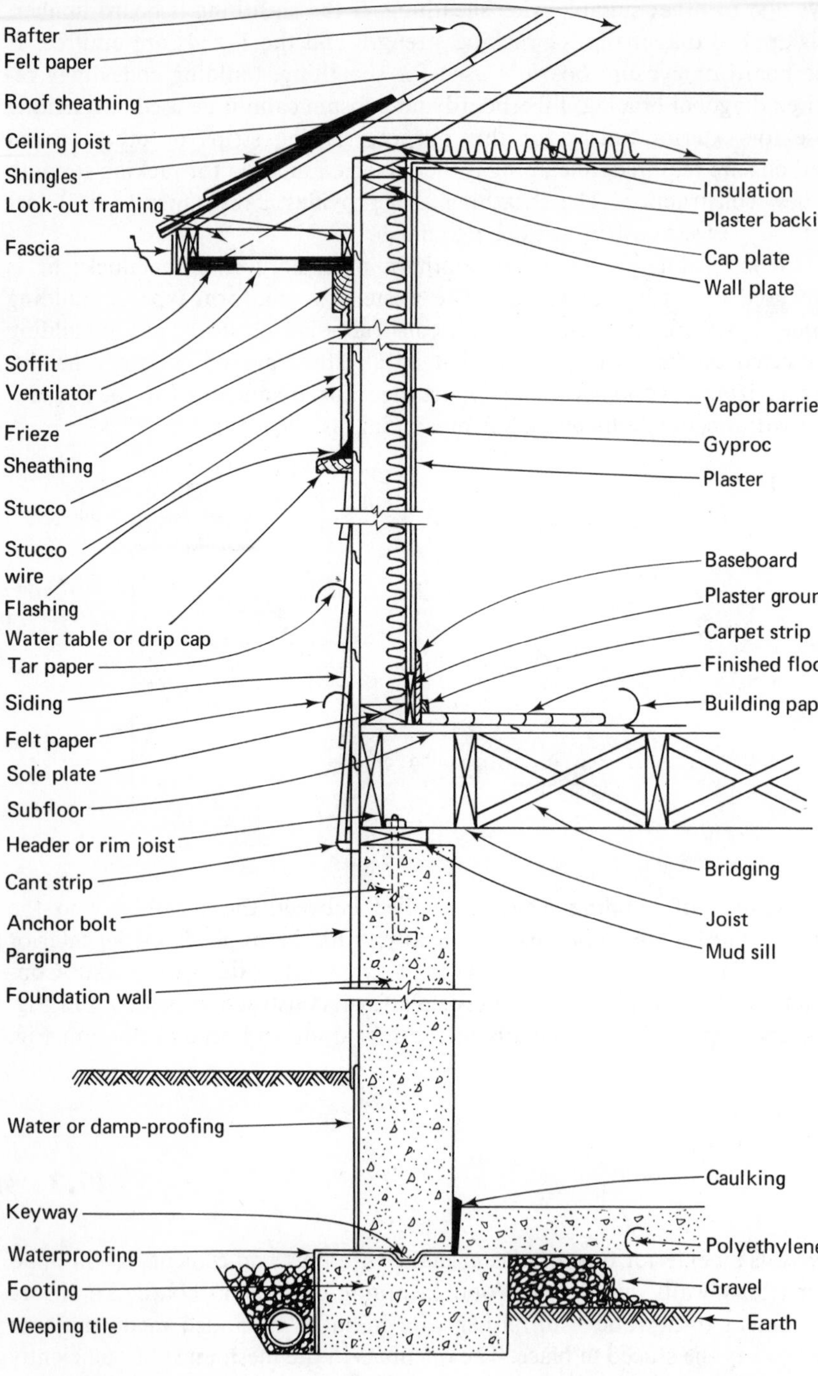
Rafter
Felt paper
Roof sheathing
Ceiling joist
Shingles
Look-out framing
Fascia
Soffit
Ventilator
Frieze
Sheathing
Stucco
Stucco
wire
Flashing
Water table or drip cap
Tar paper
Siding
Felt paper
Sole plate
Subfloor
Header or rim joist
Cant strip
Anchor bolt
Parging
Foundation wall
Water or damp-proofing
Keyway
Waterproofing
Footing
Weeping tile
Insulation
Plaster backing
Cap plate
Wall plate
Vapor barrier
Gyproc
Plaster
Baseboard
Plaster ground
Carpet strip
Finished floor
Building paper
Bridging
Joist
Mud sill
Caulking
Polyethylene
Gravel
Earth

Fig. 19-4 Wood frame construction.

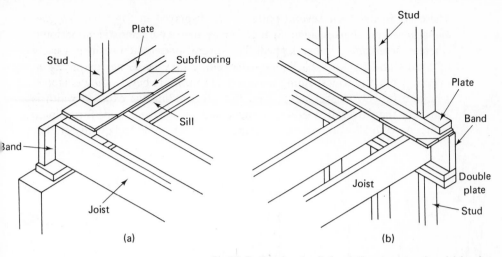

Fig. 19-5  Framing details for platform construction: (a) framing of first floor at exterior wall; (b) framing of second floor at exterior wall.

Like plaster, stucco is applied in three coats: the scratch coat, brown coat, and finish coat. The general proportioning is 1 volume of portland cement to 3–5 volumes of sand and $\frac{1}{4}$ volume of lime putty or hydrated lime. Each coat must dry before application of another coat. The finish coat is frequently colored and also textured in various styles. Colored stone may also be sprayed against the finish coat for a decorative effect.

Stucco is an economical and durable finish, but not as pleasing, popularly acceptable, or prestigious as other finishes, such as brick, stone veneer, or siding. It may be painted with stucco paints if the owner wishes to change the color.

## 19.4  siding

*Wood siding* is produced from various softwoods, preferably redwood or cedar. These two woods are relatively free of knots, straight-grained, and show minor dimensional changes with changes in moisture content.

Siding, with its horizontal shadow lines, makes an attractive exterior finish. A great many patterns are offered to the market, but these fall into three groups, distinguished by the method of jointing or lapping:

1. tongue-and-groove
2. shiplap
3. bevel siding

These types and their several patterns are displayed in Fig. 19.6. *Tongue-and-groove* and *shiplap siding* fit together by their characteristic wood joint, whereas *bevel siding* is overlapped. The several patterns come in a variety of widths and thicknesses. Square-edge siding may be as narrow as $2\frac{1}{2}$ in. overall, while the widest siding measures $11\frac{1}{2}$ in. Plain bevel siding is lapped 1 in. and rabbeted bevel siding $\frac{1}{2}$ in. For other siding the exposed width is $\frac{1}{4}$ in. less than overall width for tongue-and-groove, and $\frac{1}{2}$ in. for shiplap joints. As with all wood products, siding is available in several quality grades.

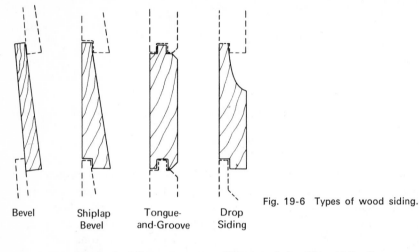

| Bevel | Shiplap Bevel | Tongue-and-Groove | Drop Siding |

Fig. 19-6  Types of wood siding.

Vertical siding types are illustrated in Fig. 19.7. Some patterns of horizontal wood siding, such as the simulated log type, can be used vertically. Siding is also used as an interior finish. Nailing methods are shown in Fig. 19.8.

Fig. 19-7  Vertical siding.

Siding is perhaps most attractive when varnished or given some other clear finish. Such finishes are much less durable than paint coverings, however. To avoid the maintenance problems of wood siding, several competitive siding materials have appeared, chiefly aluminum and rigid polyvinyl chloride.

Aluminum siding is made either 9 or 12 in. wide in 12-ft lengths, and also in wider widths for vertical siding. The sheet aluminum is about 0.02 in. thick, of a corrosion-resistant alloy such as 3003. The siding may be simply anodized, leaving the characteristic metallic luster of aluminum, or anodized and painted, or with a baked vinyl in various colors. The surface of anodized aluminum is porous so that paint and finishes will bond strongly

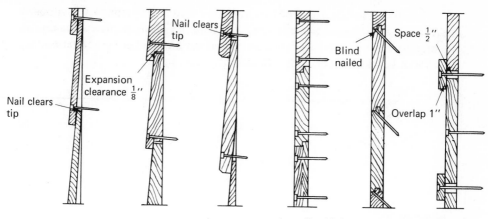

Fig. 19-8  Nailing methods for siding.

to the aluminum. Some types of aluminum siding are backed with rigid insulation.

Two styles of aluminum siding are produced: one has a slightly curved surface, the other is a flat surface (Fig. 19.9).

Fig. 19-9  Curved and flat aluminum siding.

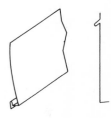

Polyvinyl chloride siding (Fig. 19.10) is produced by extruding PVC into the shape of a siding. The color is mixed into the resin before extrusion and thus is permanent.

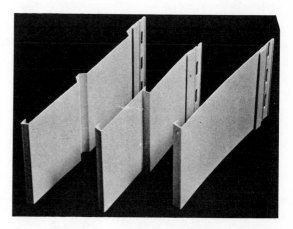

Fig. 19-10  Rigid polyvinyl chloride siding profiles.

Siding may also be made by various expedients. Tempered hardboard (Masonite) or other exterior grade of material such as plywood may be cut into strips of suitable width and applied as siding, either in the overlapping style of bevel siding or with a rabbeted wood strip (Fig. 19.11).

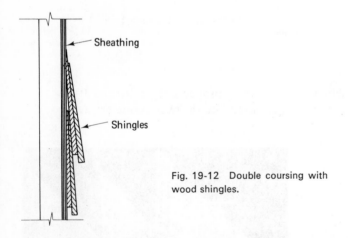

Sheathing

Hardboard siding

Rabbeted
wood strip

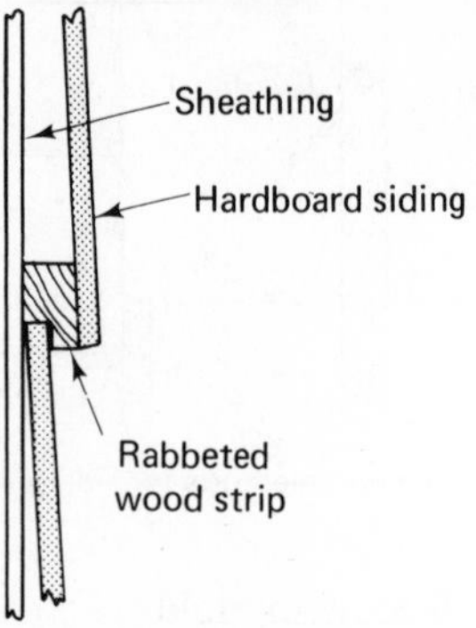

Fig. 19-11   Hardboard as siding.

Wood shingles and shakes are also employed as wall siding material. In application on vertical walls, the exposure of the shingle or shake can be increased. Double coursing is sometimes used on walls; this method adopts a double course of shingles (Fig. 19.12) to produce a heavier shadow line. The exposure may safely be increased when double shingling.

Sheathing

Shingles

Fig. 19-12   Double coursing with
wood shingles.

Asbestos–cement siding boards are 12 in. wide by 48 in. long. Asbestos siding shingles (Fig. 19.13) are 12 × 24 in. with a vertical wood-grain pattern and a wavy bottom edge. These materials are supplied predrilled for nailing, using 16-in. centers. These asbestos–cement products are applied to strapping spaced 12 in. on centers.

At the end of a wall, siding must be terminated at a corner board or suitably lapped or mitered (Fig. 19.14).

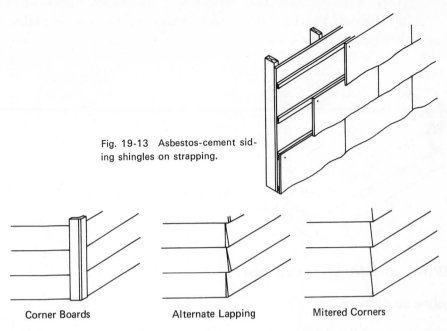

Fig. 19-13 Asbestos-cement siding shingles on strapping.

| Corner Boards | Alternate Lapping | Mitered Corners |

Fig. 19-14 Corner methods for siding.

## 19.5 brick and stone veneer

Both natural and artificial stone are used as veneer. Natural stone is used in thin veneer slabs. Artificial stone is a cast stone made by casting in molds, using either a colored mortar mix or a mix of graded mineral aggregate and a curing polyester resin. Both smooth and rough surfaces are available on artificial stone, the units being about 1 in. thick and made in various sizes up to 8 × 16 in. For installation of this veneer, a wire mesh or coarse expanded metal is applied to the wall, a layer of mortar next applied to the mesh, and the cast units set into the mortar backing. Random, course, or ashlar patterns may be used with natural stone.

Brick veneer uses any standard face brick in stretcher courses or a special thin veneer brick. These two types may be applied to the wall with the wire mesh and mortar bed as used with stone veneer. With standard brick, an alternative method is to use metal brick ties nailed to the wood sheathing of the wall as shown in Fig. 19.15.

Thin slabs of glazed and polished terra cotta (vitrolite) in various sizes and a few colors are used as an exterior veneer. These are attached with an asphalt mastic, leaving a small joint between terra cotta units to be filled with joint cement. These should be applied to wood-sheathed wall. If applied directly to a hard surface such as concrete, they can be cracked by accidental impact.

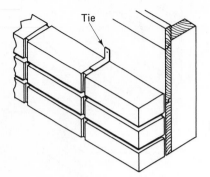

Fig. 19-15  Brick ties to wood wall.

## 19.6  board-and-batten and
## plywood finishes

A *board-and-batten* exterior finish is a finish with a vertical shadow. Boards are applied vertically to the wall and narrow strips are nailed over the vertical joints. Alternatively some pattern in cedar or redwood vertical plank siding may be used. Vertical siding uses tongue-and-groove joints.

Any exterior grade of $\frac{3}{4}$-in. plywood may be used for the boards of a board-and-batten finish. Plywood is not recommended for the battens because the sawed edge is rough and splintered. Fir plywood is not recommended because of the poor appearance of the raised grain of its summerwood.

Exterior plywood is produced in several exterior finishes for exterior wall cladding. A striated plywood with small parallel grooves is applied to the wall with the striations vertical. Other types have one surface painted with a baked smooth surface finish.

## 19.7  insulation materials
## for walls and roofs

There are eleven forms in which insulation may be provided:

    1. loose fill
    2. batts
    3. blankets
    4. block insulation
    5. structural insulation board

6. slab insulation
7. corrugated insulation
8. reflective insulation
9. sprayed insulation
10. cured-in-place insulation
11. preformed to a contour

Some of these types are not normally used in framed walls. Preformed insulation is fitted to piping and other contours. Block insulation is smaller in area than insulation board and more suited to the high-temperature requirements of mechanical equipment. Insulation that is cured in place is made from two reacting liquids which foam and solidify in the cavity into which they are pumped or sprayed. The urea–formaldehyde or urethane foams of Chapter 15 are two examples. Again, this type of insulation is not commonly used in framed walls, except when the cost of electric heat justifies the spraying of urethane foam in residences. Another type of sprayed insulation is asbestos fiber, sometimes used in residential construction.

Fibrous loose fill insulation is made from mineral, glass, or slag, which is heated and blown into thin threads. This is bagged and poured or blown into the space to be insulated. Shredded wood fill insulation is less used than in former years. Granular loose fill insulation may be of perlite or vermiculite, packed in bags of 4 ft³ and available in graded sizes. The coarser grades, Nos. 1 and 2, are used as loose fill. None of these loose fill insulations are recommended for vertical spaces because they will pack and settle with the passage of time. They are quite successful in horizontal spaces, as between ceiling joists.

Insulation batts are similar to blankets except that they are made in short lengths. Like insulation blankets, they are of mineral wool in thicknesses of 1, 1½, 2, 3, and 4 in. Widths are made to fit between studs 16, 20, or 24 in. on centers. Blankets are put up in long lengths as a roll. They are available with no covering, with paper or cardboard on one side, with wire mesh on one side, double-sided, or with reflective aluminum foil. If provided with a

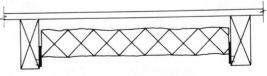

Tabs Stapled to Edge of Studs

Fig. 19-16  Stapling of blanket insulation to studs.

Tabs Stapled to Side of Studs

backing, the backing has flanges for stapling the insulation to studs in either of the methods of Fig. 19.16. Unfaced insulation batts are pressed into the stud spaces and stay in place by reason of their resilience.

The application of structural insulation board has been mentioned in its uses in roofing. These are fibrous boards of wood, cane, compressed cork, or straw, with or without a coat of asphalt. For exterior sheathing purposes, insulating board is $\frac{1}{2}$ in. thick in 4-ft widths, square edged, and impregnated with asphalt. Lengths are usually 8 ft, although longer lengths are available. As interior wallboard, such boards may be $\frac{5}{16}$, $\frac{3}{8}$, or $\frac{1}{2}$ in. thick.

Rigid slab insulation is usually stiff, polyethylene slabs are not. There are several types: mineral wool with a binder, cellular glass, foamed concrete, cellular hard rubber, shredded wood with cement, styrofoam, polyurethane, and polyethylene.

Mineral wool slabs are chiefly used for roof decking and insulation. Cellular glass is expanded glass cast into blocks and cut to size. The slabs are $12 \times 18$ in., in thicknesses of $1\frac{1}{2}$, 2, $2\frac{1}{2}$, 3, and 4 in. This is a strong material with a density of 9 lb per ft$^3$. It is applied to walls or roofs with asphalt emulsion or hot asphalt.

Foamed concrete has a density of about 40 lb per ft$^3$. This material is made into insulating structural precast roof slabs for decking. No further insulating material is required for such a roof.

Cellular hard rubber is foamed with nitrogen and available in the usual range of thicknesses for slab insulation. Wood fiber makes a rigid insulation when formed with portland cement. The author installed an interesting rigid type of insulating slab in the Canadian north about 20 years ago; it was a mixture of Atlantic seaweed and cement.

The foamed plastics have been previously discussed.

The various types of slab insulation are used for roof insulation, perimeter insulation, subbase insulation, wall insulation, and insulation in the cavity of cavity walls. Subgrade installations must use insulating slabs that are impervious to moisture. Asphalt emulsions, synthetic adhesives, and portland cement mortars are used to bond the slabs to masonry walls, the choice of adhesive being determined by the type of insulation.

Polished aluminum foil reflective insulation is usually furnished bonded to paper, corrugated paper insulation, mineral wool blankets, or wallboard. The function of aluminum is to reflect radiant heat back to its source; to be effective the reflective surface must reflect back to an air space at least $\frac{1}{2}$ in. wide. The reflective surface must face the heat source or hot side of the building. The sheets of foil are installed vertically between studs, or, if 36-in. widths are used, the sheet will span two stud spaces.

Paper honeycomb is produced as an insulating partition material (Fig. 19.17). The paper honeycomb core, 1–4 in. thick, is stiffened with paper faces that are sprayed with a light coating of portland cement. This honeycomb sandwich is used for non-load-bearing partitions. The sides are finished with plaster.

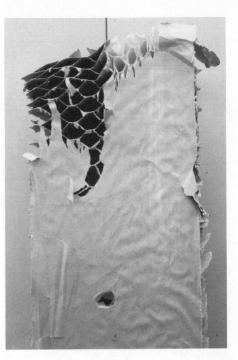

Fig. 19-17 Corrugated paper insulating board.

## 19.8 insulating foam as a structural wall material

In this section two promising and interesting developments in house construction are described. Both use foamed plastics.

In Chapter 15 it was noted that foamed polyurethane is not normally economical for house insulation except under special circumstances. The polyurethane resins currently cost 50 cents per pound as supplied in 55-gal drums, which is $8\frac{1}{3}$ cents per board foot for 2-lb foam if there are no losses due to overspray and no spraying costs. Installation costs would be onerous for a spray job comprising a single house.

Nevertheless, polyurethane was extensively used in 1970 in Heritage Village in Connecticut near New York City. The insulation originally selected was fiberglass; this was replaced by urethane foam in later houses in the development. The wall construction for the houses was modified by using the foam to stiffen the wall frames against racking, thus eliminating a $\frac{1}{2}$-in. fiberboard sheathing from the wall construction. The altered wall construction is shown in Fig. 19.18. The revised construction details received local building code approval.

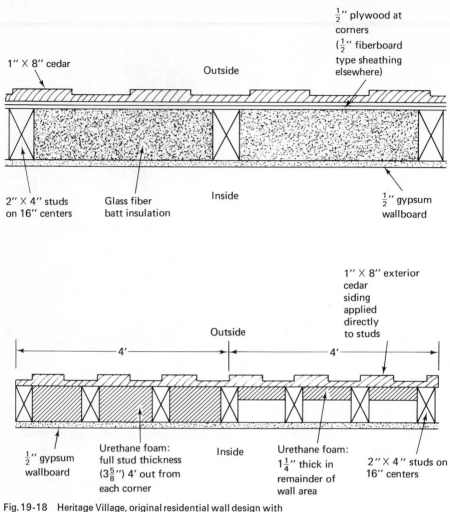

Fig. 19-18 Heritage Village, original residential wall design with glass fiber insulation; and the later design using polyurethane foam.

The other case to be mentioned uses foamed polystrene block wall, a design by Structural Foam, Inc., of Parker Ford, Pennsylvania. The structural foam wall is built from expanded polystyrene blocks of 1 pcf density. The blocks are 6 in. wide by $11\frac{3}{4}$ in. high, in any lengths. The blocks fit by a tongue-and-groove type of joints. Large holes run vertically through the tiers of blocks, and concrete reinforced with rebar is poured into these holes. The foam serves as a wall, an insulation, and a concrete form. At the top of the wall a reinforced bond beam is poured to carry the roof and bond the structure together. Aluminum or vinyl siding is used for exterior finish, and window and door frames are of rigid polyvinyl chloride. The interior finish may be panels, drywall, or gypsum plaster.

## 19.9 concrete block

## and brick walls

The cheapest types of industrial buildings are the prefabricated steel building (Fig. 19.19) and the concrete block wall structure (Fig. 19.20). The prefabricated steel building has a steel skeleton to carry the building weight and

Fig. 19-19   Research house designed by the Steel Company of Canada, Limited. Floor joists, studs, roof trusses, roofing panels, window and door frames, house siding, and even the fence in the photograph are of steel. The average home contains 1 ton of steel; this house contains $8\frac{1}{2}$ tons.

imposed loads. Its walls are of the curtain-wall type, exterior sheet steel panels galvanized, sheet steel insulated roof, wall insulation of mineral wool batts or blankets between the outer steel skin and an interior wall skin of 30-gage galvanized steel. The concrete block wall is of the bearing-wall type.

The allowable thickness of concrete block walls is governed by the prevailing building code. For a residential building, usually the following thickness specifications apply to concrete block exterior walls:

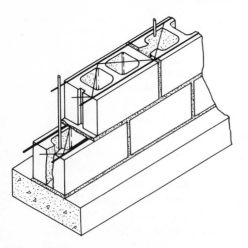

Fig. 19-20 Concrete block construction using a patterned block.

| No. of Stories | Basement | First Story | Second Story | Third Story |
|:---:|:---:|:---:|:---:|:---:|
| 1 | 12 (or 8) | 8 (or 6) | | |
| 2 | 12 (or 8) | 8 | 8 | |
| 3 | 12 (or 8) | 8 | 8 | 8 |

For buildings other than residential, exterior block walls are required to be 12 in. wide for basement, first story, and second story, but 8 in. may be allowed for single-story buildings.

Details for a reinforced concrete masonry foundation wall are suggested in Fig. 19.21. Here reinforced masonry with horizontal joint reinforcement

Fig. 19-21 Block foundation wall with joint reinforcement.

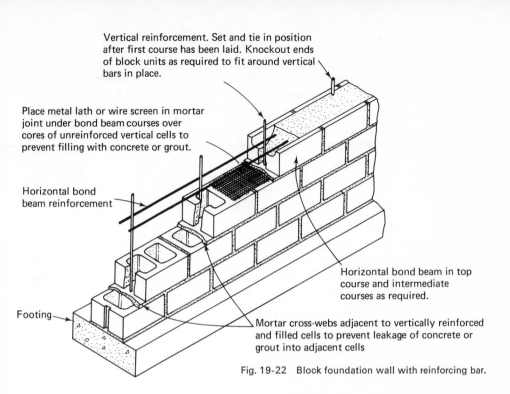

Vertical reinforcement. Set and tie in position after first course has been laid. Knockout ends of block units as required to fit around vertical bars in place.

Place metal lath or wire screen in mortar joint under bond beam courses over cores of unreinforced vertical cells to prevent filling with concrete or grout.

Horizontal bond beam reinforcement

Footing

Horizontal bond beam in top course and intermediate courses as required.

Mortar cross-webs adjacent to vertically reinforced and filled cells to prevent leakage of concrete or grout into adjacent cells

Fig. 19-22   Block foundation wall with reinforcing bar.

is used. Alternatively the method of Fig. 19.22 may be used, with reinforcing bar instead of martar joint reinforcement. A detail of the galvanized wire joint reinforcement is shown in Fig. 19.23. Figure 19.24 details a joint between a poured or precast concrete floor and a foundation wall.

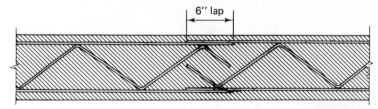

6" lap

Fig. 19-23   Splicing method for joint reinforcement.

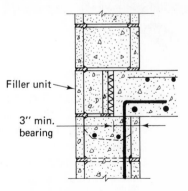

Filler unit

3" min. bearing

Fig. 19-24   Joint between a block foundation and a poured concrete floor.

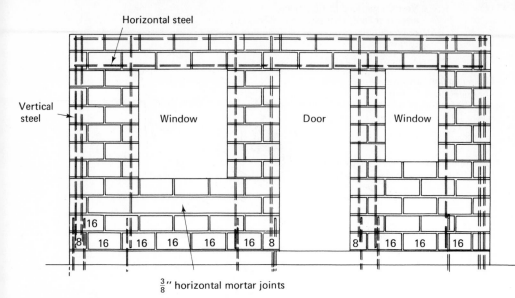

Horizontal steel

Vertical steel

Window

Door

Window

16

8 | 16 | 16 | 16 | 16 | 16 | 8

8 | 16 | 16 | 16

$\frac{3}{8}''$ horizontal mortar joints

Fig. 19-25 Eight-inch modular wall layout showing vertical and horizontal reinforcement. Horizontal reinforcement is required at the top of footings and. at the top of wall openings.

Layup of a typical wall is given in Fig. 19.25, showing also a bond beam with reinforcing rods. Joint reinforcement is shown in Fig. 19.26,

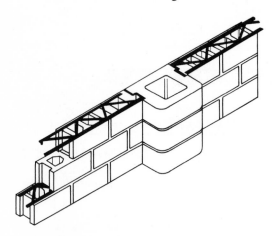

Fig. 19-26 Pilaster and joint reinforcement.

with a pilaster. Corner construction is shown in Fig. 19.27, lintel construction in Fig. 19.28. A detail of a complete wall with reinforcing bar is displayed in Fig. 19.29. Alternatively, joint wire reinforcing may be used every second course.

Various types of masonry wall ties are illustrated in Fig. 19.30. These are of various designs to serve various circumstances, such as cavity walls

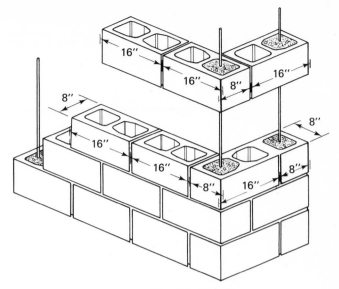

8" wall to 8" wall

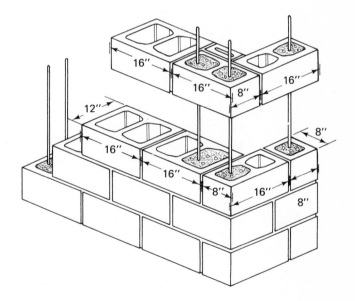

12" wall to 8" wall

Fig. 19-27   Corner construction.

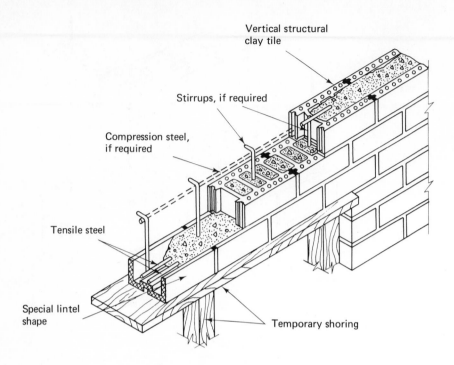

Vertical structural
clay tile

Stirrups, if required

Compression steel,
if required

Tensile steel

Special lintel
shape

Temporary shoring

Fig. 19-28   Lintel construction using U-shaped lintel blocks.
There are several other methods of lintel construction.

(usually with a 2-in. cavity) or backup masonry. The adjustable wall ties, made in two pieces of galvanized wire, are used when the brickmason wishes to raise one wythe before building up the other, or to adapt to construction where courses do not align. The adjustable wall tie will not properly bond two wythes unless the eye on the female piece is positioned close to the face of the brickwork (Fig. 19.31). The other or male piece of the adjustable wall tie is called the pintle piece. For the required use and spacing of wall ties, consult local building codes.

## 19.10   ceramic veneer

## and stone facing

Various combinations are possible with veneered types of masonry walls. Structural tile, concrete block, or common brick may back up face brick, facing tile, ceramic veneer, terra-cotta, or stone.

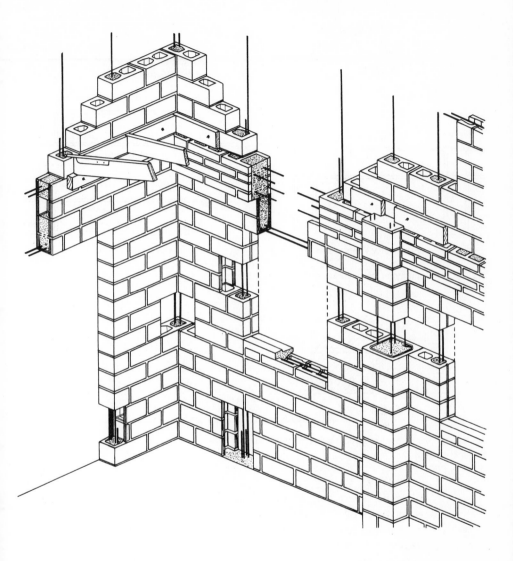

Fig. 19-29   Complete wall detail showing a pilaster and re-inforcing.

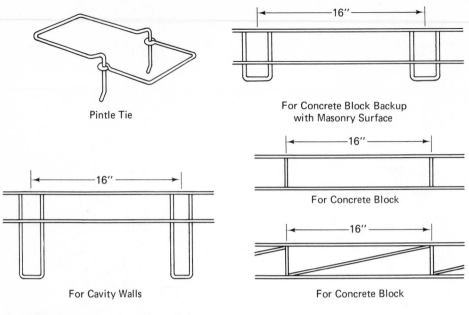

Pintle Tie

For Concrete Block Backup
with Masonry Surface

For Concrete Block

For Cavity Walls

For Concrete Block

Fig. 19-30   Some types of masonry wall ties.

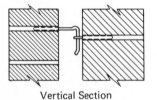

Vertical Section

Fig. 19-31   Installation of double pintle tie, with the eye close
to the brickwork.

*Ceramic veneer* may be of two application types: adhesion or anchor. Adhesion types of this veneer are about 1 in. thick and are bonded to the wall with a wall coat of mortar. Anchor types of such veneer are attached to the backup wall by metal anchors (Fig. 19.32). The latter method of application permits the use of thicker veneer, sizes ranging from $1\frac{1}{2}$ to 3 in. The usual face area is 18 × 24 in. or smaller.

For installing *stone veneer*, metal anchors must be employed (Fig. 19.33). Stone veneer does not usually exceed about 5 in. in thickness, although bond stones of greater thickness may be used for bonding to a masonry wall. A bond stone does not require metal anchors. The number of anchors required depends on the size of the stone slab:

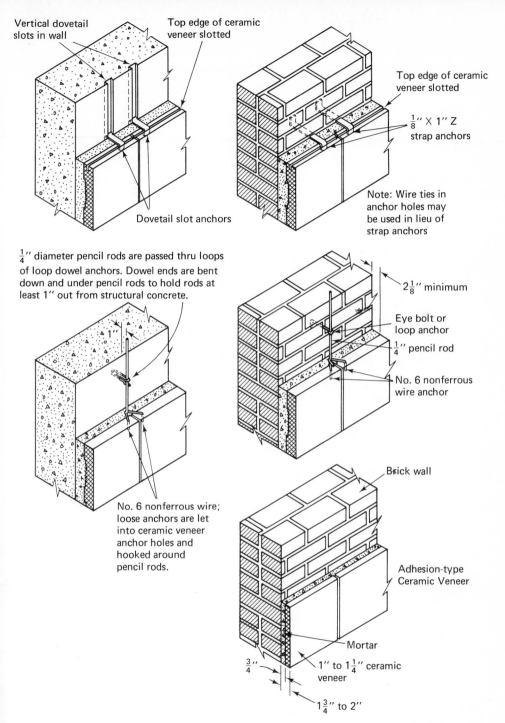

Fig. 19-32 Anchoring ceramic veneer. The adhesion type may also be laid over wood sheathing and metal mesh.

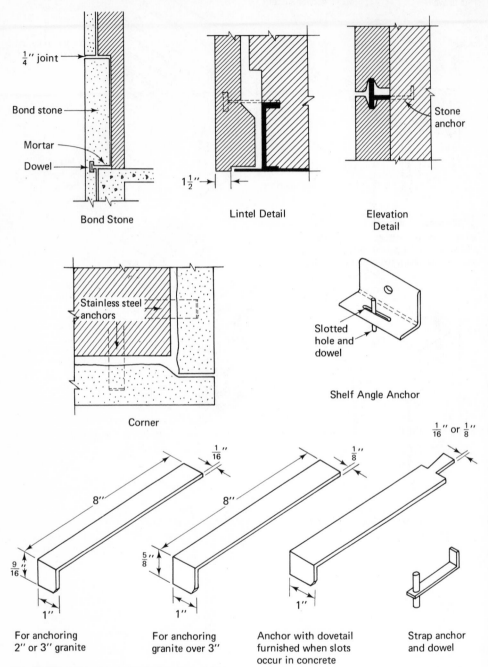

Fig. 19-33 Anchoring methods for stone veneer.

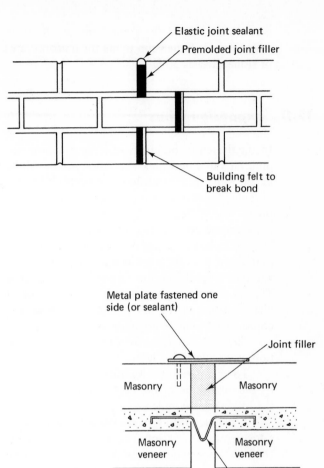

Fig. 19-34   Typical expansion joints.

| Size of Stone Veneer (ft$^2$) | Number of Anchors |
|---|---|
| up to 2 | 2 |
| 2–4 | 3 |
| 4–12 | 4 |
| 12–20 | 6 |
| over 20 | 1 extra anchor for every additional 3 ft$^2$ |

Concrete or masonry backup material should be damp-proofed with asphalt before the stone is set in place. Moisture must not be allowed to accumulate behind the stone, and an air space to allow circulation of air may be used. Expansion joints for masonry are usually needed if the length of stone exceeds 30 ft.

## 19.11   expansion joints

In the design of buildings of large dimensions and complex shape, such as multiple building wings, expansion joints will be required. If thermal strains in the walls, roof, and floors of such buildings cannot be relieved, stress is developed, causing distortion and cracks. Such expansion joints can sometimes pose difficult design problems, and rather severe demands may be made on the expansion-joint material. In a building with several wings, each wing must be independent in its thermal and other movements, which means that the expansion joint must section the complete building cross section without prejudicing the strength of the structure or its watertightness. Typical expansion joint details are shown in Fig. 19.34. Note the use of a filler material in the joint, finished with an elastomeric sealant and a soft copper waterstop where necessary.

Thermal expansion is a major preoccupation in curtain-wall construction, particularly for large areas of glazing.

To calculate total thermal expansion, multiply the temperature change by the dimension of the part and the coefficient of thermal expansion as in the following example.

A plastic glazing panel has a vertical dimension of 9 ft. It is subject to a temperature variation from 0 to 100°F. The plastic material has a thermal coefficient of expansion of 0.00005 in./in./°F. Determine the probable amount of expansion over this temperature range.

Convert 9 ft to 108 in. The thermal expansion = 108 in × 100° × 0.00005 = 0.54 in. In designing a frame for this glazing panel, an expansion allowance of somewhat more than this amount must be provided in the design.

### COEFFICIENTS OF LINEAR THERMAL EXPANSION

(units of inches per inch per °F)

| acrylic | 0.00005 | limestone | 0.000004 |
|---|---|---|---|
| aluminum | 0.000012 | marble | 0.000006 |
| brick | 0.0000031 | plaster | 0.000009 |
| clay tile | 0.0000035 | polyvinyl chloride | 0.00003 |
| concrete | 0.0000065 | stainless steel | 0.000009 |
| copper | 0.000009 | structural steel | 0.0000065 |
| glass | 0.000005 | wood (across grain) | 0.000002 |
| granite | 0.000004 | wood (parallel to grain) | 0.000003 |

Fig. 19-35  Curtain-wall construction : the 57-story Commerce Court West in Toronto, Canada. The curtain walls are glass and stainless steel 304.

## 19.12 curtain-wall construction

In *curtain-wall construction* (Fig. 19.35) the walls do not carry the weight of the stories and roof above. The weight of the building is carried on an internal skeleton of steel or reinforced concrete, the walls serving only as an envelope to enclose the building. The curtain wall is a challenging concept and has encouraged the use of new materials and sandwich components.

Fig. 19-36  Basic types of curtain-wall construction.

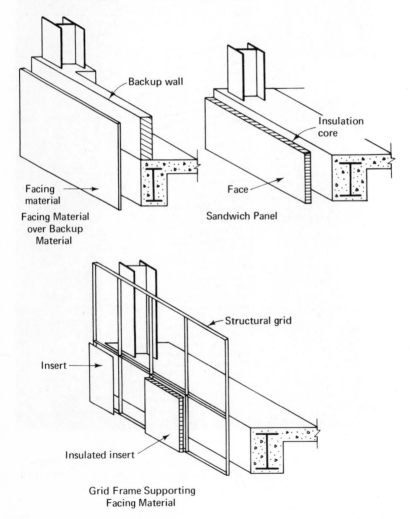

Facing Material
over Backup
Material

Sandwich Panel

Grid Frame Supporting
Facing Material

But all new techniques disclose new problems to be solved, and so has it been with curtain walls.

The first curtain-wall systems used large sheets of glass between the structural beams. Traditional glazing compounds as used for domestic window glazing were applied to seal these large sheets of glass. Such standard glazing materials harden with age and can safely close up only very minor amounts of deflection. The large curtains of glass were found to deflect significantly from wind pressure, vibration, and thermal movements because of their large size and large temperature variations. The caulking problem became a very serious one, and although caulking compounds of outstanding performance are now available, perhaps none is as good as we might wish.

Three basic types of curtain wall may be distinguished:

1. Curtain-wall backup construction with facing material.
2. Prefabricated sandwich panel curtain wall.
3. Grid frame or subframe curtain wall.

In the first type, the envelope of the building may be closed with insulating block, hollow tile, or other material, to provide the required thermal insulation and fire rating. A skin or veneer of exterior finish for appearance and watertightness is applied to the backup material (Fig. 19.36).

The panel type of wall uses large panels made of an insulating core covered with an interior and an exterior facing material. No backup material is used, provided that the panels have the required stiffness (Fig. 19.37).

The grid frame curtain wall (Fig. 19.38) is so named because the structure of the wall is a grid framework of aluminum extrusions or of light steel

Fig. 19-37 Typical curtain-wall panel construction.

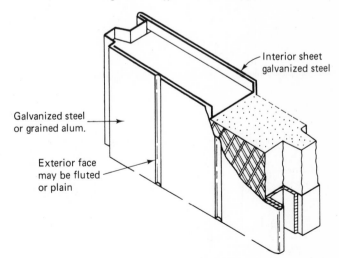

Interior sheet
galvanized steel

Galvanized steel
or grained alum.

Exterior face
may be fluted
or plain

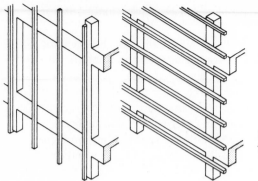

Fig. 19-38 Curtain-wall grid frames.

sections. This supporting grid is closed in with glazing material, insulated sandwich panels, or other suitable panels which provide fire rating and insulation. Some anchoring methods for these subframes are indicated in Fig. 19.39.

## 19.13 masonry

*Masonry walls* may be used in curtain-wall construction, the weight of each wall of the story being carried on steel or concrete beams. The familiar

Fig. 19-39 Anchoring methods for grid frames.

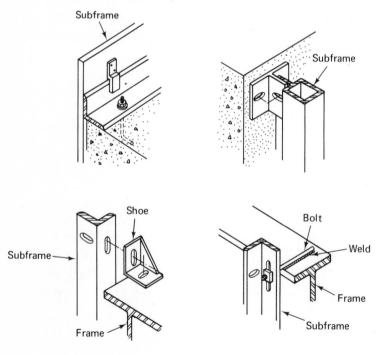

combinations of common brick for backup and face brick veneering, or of tile backup and ceramic tile or stone veneer, and other combinations are all possible in curtain walls.

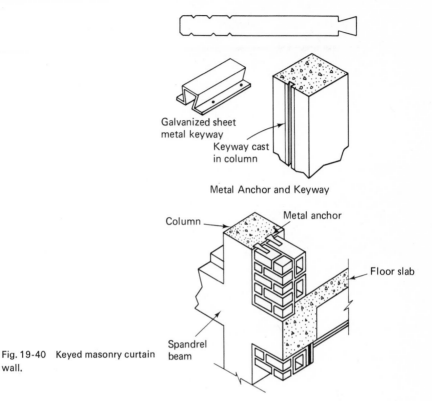

Metal Anchor and Keyway

Fig. 19-40  Keyed masonry curtain wall.

If the building has a reinforced concrete structural frame, metal ties are used to anchor the ends of the masonry wall to the concrete columns. Such ties are flat steel sheet, one end keyed to fit a matching keyway in the concrete column and the other laid into a brick mortar joint, as in Fig. 19.40. The keyway in the column is lined with a formed galvanized steel lining. If a steel building column is fireproofed with concrete, the keyway may be cast into the fireproofing concrete.

If the building has a steel frame, the masonry of the curtain wall may enclose the columns, or the columns may be exposed (Fig. 19.41). The masonry may be carried directly on the spandrel beams or on a lintel angle.

Both adhesion and anchor types of ceramic veneer are used as curtain-wall facing materials. The anchored ceramic veneer (Fig. 19.42) will also be carried on steel angles.

In curtain-wall construction, two types of stone veneer are employed, either a stone veneer or a sandwich of stone backed with insulation. Stone veneer is applied in any of the methods discussed in Section 19.10 (Fig. 19.43).

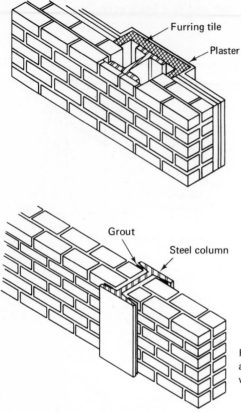

Furring tile

Plaster

Grout

Steel column

Fig. 19-41 Steel frame concealed and exposed by masonry curtain wall.

Fig. 19-42 Anchoring ceramic veneer.

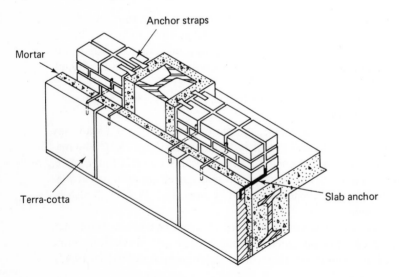

Anchor straps

Mortar

Terra-cotta

Slab anchor

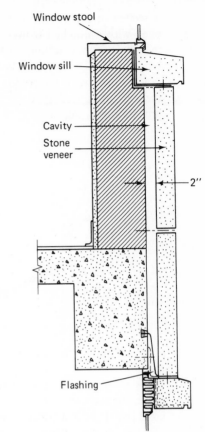

Window stool

Window sill

Cavity

Stone
veneer

2"

Fig. 19.43 Anchoring stone
veneer.

Flashing

Sandwich panels are made of a thin stone face about 2 in. thick backed
with a rigid type of insulation such as styrofoam or fiberboard or of fiberglass
insulation enclosed in sheet steel.

## 19.14 concrete

Concrete is applied to curtain walls as precast panels, which may be made
either of a standard or a lightweight concrete. Insulation may be added to
the prefabricated panel as backing.

Plain concrete has a somewhat unattractive appearance and color, and
some type of texturing or surface is usually given to the concrete panel.
A layer of white cement and colored aggregate or a ceramic veneer may be
applied. More recently exposed aggregates using thermosetting epoxy resin
or a latex as binder have appeared. Exposed aggregates are of two types:
stone-seeded and trowled mixtures. In the stone-seeded type the aggregate

is seeded or forced into the binder or matrix. The epoxy binder is first applied with a notched trowel. In the other type, the aggregate is mixed with the binder for troweling on the surface. The aggregate is exposed after the binder dries. The latex mixes are designed to be mixed with cement and silica sand to produce a matrix for the seeded aggregate. Colors may be added on the job, and any desired decorative stone or chip may be bonded. Such aggregate mixes may be applied over any sound solid surface, including plywood.

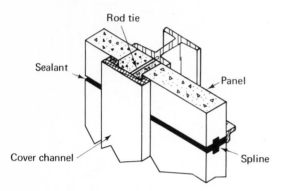

Fig. 19-44 Cover channel with curtain-wall panels.

Concrete panels are anchored to the building by any of the anchoring methods used with stone panels. The panels may have tongue-and-groove edges for joining to the adjacent panel or plain edges. Steel or aluminum cover channels (Fig. 19.44) may be used to enclose the seam where the concrete panel abuts against a building column.

Corrugated asbestos–cement board is also used in curtain-wall construction, often laminated with a rigid insulation.


## 19.15   metal sandwich panels

Steel, stainless steel, and aluminum are employed in metal curtain-wall panels. Steel and stainless steel skin panels without backing materials are

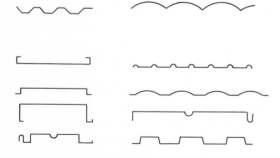

Fig. 19-45   Typical shapes of steel skin panels.

roll-formed for stiffness in such shapes as illustrated in Fig. 19.45; aluminum skins may be roll-formed or extruded, and may be anodized for improved surface protection (Fig. 19.46). Steel panels are available surfaced with porcelain enamel.

Fig. 19-46   Acrylic-coated aluminum curtain wall.

Sandwich panels with metal facing have an insulating core of fiberglass or styrofoam. Some sandwiches include a stiffening of wood hardboard or asbestos–cement board bonded to the metal skin. Insulation thickness is either $1\frac{1}{2}$ or 2 in., giving a $U$ factor for the panel of about 0.15. Most of these metal sandwich panels weigh about 7 psf. Figure 19.47 illustrates some types of construction.

## 19.16   glass curtain walls

Besides its use as large glazing panels, glass is used in curtain-wall construction in both block and tile form. Glass block is 4 in. thick, in 6 × 6, 8 × 8,

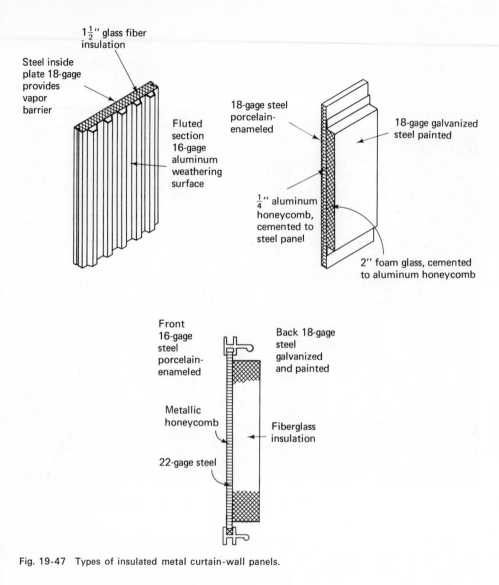

Fig. 19-47 Types of insulated metal curtain-wall panels.

and 12 × 12 sizes, with a half size 4 × 12 in. Glass tile are made up in larger panels in an extruded aluminum perimeter frame with neoprene or other types of gaskets to seal the glass panel. The frames are joined together with batten strips to cover the joints. Glass blocks also require a resilient mounting. They are laid in stack bond on a resilient base, with a fiber pad against the jamb, as in Fig. 19.48.

Insulating glass panels are made by mounting two sheets of plate glass together with an air space between them. The air space must be dehydrated and hermetically sealed if condensation on the inside surfaces of the glass sheets is to be prevented.

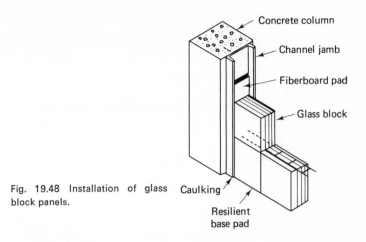

Fig. 19.48 Installation of glass block panels.

Concrete column

Channel jamb

Fiberboard pad

Glass block

Caulking

Resilient base pad

Walls may also be built of window walls, which are windows with wood or metal sash and joined in long lengths to cover the wall.

Plate glass $\frac{1}{4}$ in. thick is a familiar type of curtain-wall material on the fronts of commercial buildings. These large sheets of plate glass may have a maximum dimension of 144 in. Glass with a backing surface of procelain enamel or vinyl is also produced. Such opaque backings result in greater thermal expansion in the glass. Clear glass transmits most of the infrared radiation that falls on it from the sun. An opaque backing absorbs this radiation, thus raising the temperature of the glass. Such glass panels must have resilient mountings that are designed with some care.

## 19.17 plastics

The use of foamed plastic insulation in curtain-wall sandwich panels has been discussed throughout this chapter. Polyurethane is self-bonding when foamed. Despite its slightly higher cost than styrofoam, this self-bonding property accounts very often for its selection instead of styrofoam, which requires an adhesive or some mechanical method of attachment or retainment.

Plastics are also used as curtain-wall surface material. Vinyl-coated steel is commonly found on prefabricated steel buildings. Fiberglass-reinforced polyester is also used, more usually for multistory types of curtain-wall construction. For stability against ultraviolet degredation, the polyester may include a chemical stabilizer or may be coated with an acrylic film. Such panels may be clear, colored, or translucent.

Plastic curtain-wall panels usually employ a grid frame (subframe) type of curtain-wall construction.

Fig. 19-49  Two examples of artistic interior finish, brick mural
and a mural in mosaic, both portraying building trades.

### 19.18  *wood finishes*

Many materials used for exterior wall finishes are also suitable for interiors: brick or mosaic, for example (Fig. 19.49).

Wood covering for interior walls may be applied as boards or sheets, including plywoods and composition boards. Wood boards for interior finish are selected for color, texture, or pattern, such as knotty pine. Plywood paneling may be of softwood or hardwood, with the surface veneer selected for interesting grain patterns or colors. Douglas fir, although it has a coarse pattern of early and late wood, can be treated by wire brushing or etching to produce a more attractive surface. Etching removes the softer earlywood. Hardboards, fiberboards, and other boards may be made into attractive and economical interior paneling by printing a wood grain on a surface film.

Fig. 19-50  Some commonly used wood moldings.

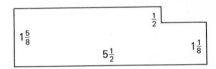

Door Jamb 2103

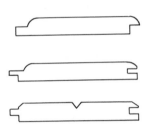

Paneling

Quarter-round

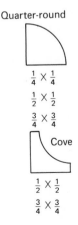

$\frac{1}{4} \times \frac{1}{4}$

$\frac{1}{2} \times \frac{1}{2}$

$\frac{3}{4} \times \frac{3}{4}$

Cove

$\frac{1}{2} \times \frac{1}{2}$

$\frac{3}{4} \times \frac{3}{4}$

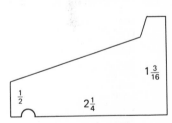

Water Table 2174

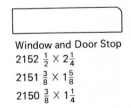

Window and Door Stop

2152 $\frac{1}{2} \times 2\frac{1}{4}$

2151 $\frac{3}{8} \times 1\frac{5}{8}$

2150 $\frac{3}{8} \times 1\frac{1}{4}$

Irregularly spaced vee grooves are cut into the face to simulate a random planking effect. Other hardboards have a baked enamel coating, and are applied in 4-in. squares to simulate ceramic wall tile. Hardwood plywood favored for interior work includes mahogany, walnut, birch, beech, oak, ash, and teak.

Wood trim is usually made from fir or mahogany. Such trim includes baseboard molding, cove molding, door jamb, half-round, and a wide range of other shapes (Fig. 19.50). Because of their diversity, such trim items are usually specified by number, since identification by name may be uncertain. Some of these wood trim shapes are being replaced by extruded rigid foamed polyvinyl chloride, which has a striking likeness to wood. The use of rigid PVC in window and door frames is increasing. The outstanding advantage of such plastic trim is that painting or repainting is not required, although if a color change is desired, paint will adhere to it.

## 19.19 gypsum and plaster walls

Gypsum board is also used as interior wallboard. Some types are produced with a surface painted with a simulated wood grain. Gypsum drywall board is nailed up, the joints filled and taped, and a paint or a plaster finish applied. The plaster finish may be smooth, stippled, or sand finished. These finishes may use acoustical plaster for sound insulation. Plaster is often applied by spraying. Gypsum and plaster finishes are suited to both walls and ceilings.

## 19.20 various wall treatments

Virtually all the clay products used in exterior finish may be used on interior walls. Lightweight concrete block when painted makes an attractive and economical wall. Ceramic wall tile and ceramic mosaic are used in wet areas such as kitchens, bathrooms, and washrooms, frequently only on the lower half of the wall. The same wall treatment is often used in other locations where its appearance or cleanliness is desired. Porcelain-enameled steel wall tile is available in a range of colors to simulate ceramic wall tile.

Stone veneer is usually found only in the interior of prestige buildings and office areas.

## 19.21 plastics

Plastic wall tile of polystyrene or urea–fomaldehyde are made in a range of colors. Plastic wall fabrics are applied to walls in a way similar to the application of wallpaper.

Fig. 19-51 Stainless steel panels on an interior column.

Plastic laminates (Arborite, Micarta, Formica) are popular for counter-tops and table tops, and are sometimes used as wall covering. The structure of these laminates is explained by Fig. 19.52. The base is a multiple-layered kraft paper bonded with phenol–formaldehyde. This is covered with a color-patterned sheet saturated with melamine–formaldehyde, chosen for its ability to give excellent color effects. A top layer of melamine–formaldehyde

Fig. 19-52 Plastic laminate for counter tops.

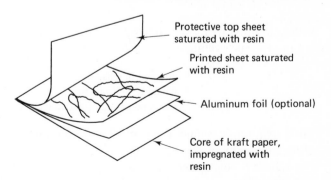

Protective top sheet saturated with resin

Printed sheet saturated with resin

Aluminum foil (optional)

Core of kraft paper, impregnated with resin

is imposed. For resistance to cigarette burns, a sheet of aluminum foil may be interposed between the base and decorative layers. Because of its high thermal conductivity the aluminum dissipates any heat from cigarettes or hot cooking pots.

## 19.22   ceiling tile and strip

There are many types of ceiling tile. Either a mineral fiber or a wood fiber is used, and the tile may be designed for acoustic treatment of the room. Acoustic treatment is not usually needed in residential construction. Butt edges or tongue-and-groove edges are supplied and the tiles are prepainted

Fig. 19-53   Suspended acoustic ceiling.

with a white vinyl finish. Dimensions are $12 \times 12$ in., although larger sizes are in use. Such ceiling tile may be mounted by three methods:

1. adhered to plasterboard
2. nailed to furring strips
3. mounted on a suspended ceiling

These lightweight ceiling tiles are highly porous and cannot be used in areas of high humidity or condensation.

Linear ceiling systems use long strips of anodized aluminum hung on a suspended ceiling (Fig. 19.54). The aluminum strips are perforated with fine holes to provide sound absorption. The anodized surface of the aluminum is an excellent paint base. The enamel finish is applied at the factory, however. This type of ceiling strip has a 4-in. module, so lighting and ventilating strips can be incorporated into the ceiling by using the same module.

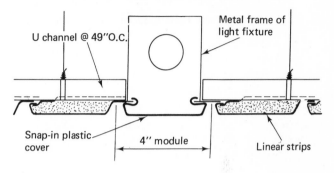

U channel @ 49"O.C.

Metal frame of
light fixture

Snap-in plastic
cover

4" module

Linear strips

Fig. 19-54   Linear ceiling.

## PERTINENT ASTM SPECIFICATIONS

C578-65T (Part 14) Preformed Cellular Polystyrene Thermal Insulation
C591-66T (Part 14) Rigid Preformed Cellular Urethane Thermal Insulation
C553-64 (Part 14) Mineral Fiber Blanket and Felt Insulation

## QUESTIONS

1   What is a curtain wall?

2   Why is wire needed with stucco?

3   What advantages does aluminum siding offer?

4   What is a bond stone?

5   Differentiate between a batt and a blanket of insulation.

6   Why is the vapor barrier placed on the warm side of a wall?

7   What is a party wall?

8   What is the principle of the platform type of wall frame?

9   What is the principle of reflective insulation?

10   Why would steel foil not be used for reflective insulation?

**11** What is meant by perimeter insulation?

**12** Define a cavity wall and a wythe.

**13** When are adjustable wall ties for brickwork employed?

**14** A brick building 100 ft long has a decorative PVC strip running its full length. The building was erected when the temperature was 60°F. At a temperature of 100°F, how much longer is the decorative strip than the building wall?

**15** List the three basic types of curtain-wall construction.

Many functions are required of a floor. It must first be structurally strong to bear the floor loadings to be imposed, level, easy to maintain, wear resistant, and resistant to moisture transmission. Secondary requirements may include low noise generation and transmission, resilience, color, and pattern. Special characteristics may also be required, such as low electrostatic generation in hospitals, or resistance to chemical attack. The cost of the floor is an important characteristic, of course, and includes the three elements of materials, installation, and maintenance.

## 20.1 subfloor

A *subfloor* is an unfinished floor that serves as a base for the finished floor surface. The subfloor must support the floor loads without excessive deflection. It is particularly important for the subfloor to be smooth, since any irregularities of surface will be reflected in the finish floor laid above it. Moisture resistance of the subfloor prevents loosening or degradation of the finished flooring.

# FLOOR

# 20

Concrete and wood are the most common subfloor materials. Wood subfloors must be level, stiff, and nailable. Usually 1-in. floor boards or plywood are used to surface a subfloor. If 1-in. board is used, it may be laid at right angles or diagonally to the wood joists.

Fig. 20-1   Lightweight concrete roof and floor plank.

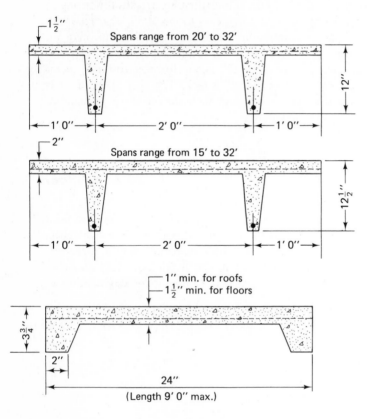

Fig. 20-2   Reinforced concrete double-tee units and channel plank for roofs and floors.

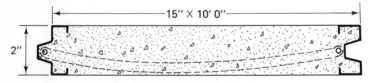

Lightweight Concrete Plank

Steel-edge lightweight aerated concrete plank integrally cast into galvanized edgings of tongue-and-groove on all floor sides. Weight 13 psf

The properties of poured concrete are such as to make it an exceedingly satisfactory material for subfloor construction. It is monolithic, strong, fire resistant, and economical. It can be installed below grade without being harmed by moisture. Instead of a poured concrete subfloor, lightweight concrete plank, either with plain sides or tongue-and-groove sides as in Fig. 20.1, may be installed. Other types of concrete floor systems include the reinforced channel plank and tee plank of Fig. 20.2, which may be laid on top of steel beams.

Cellular steel beams are another type of subfloor construction, shown in Fig. 20.3. This subfloor structure, carried on steel beams, may carry 2 × 4 wood sleepers on which a finished floor of wood is laid, or be made of poured concrete with or without a top finish. The long cells in this type of construction are often useful for stringing pipe and electrical conduit.

In selecting a finished floor material, a major consideration is the floor level with respect to grade: below-grade, on-grade, or above grade.

Fig. 20-3   Wood and concrete floors on cellular steel beams.

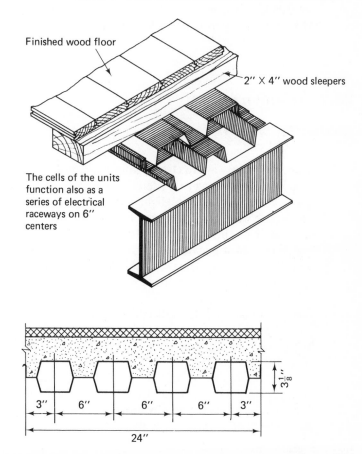

Finished wood floor

2″ × 4″ wood sleepers

The cells of the units function also as a series of electrical raceways on 6″ centers

3″   6″   6″   6″   3″

$3\frac{1}{8}$″

24″

Any floor below the ground level is a below-grade floor. Moisture from the ground can pass through concrete subfloors. This moisture can carry with it dissolved alkaline salts that are universally present in clays. Both the moisture and its salts may destroy the adhesive bond used to attach some finish floors. To prevent such damage, a 6-mil thickness of polyethylene film is laid before the concrete subfloor is poured. Such a membrane is also required for on-grade construction.

Several types of resilient tile flooring are acceptable for installation below grade. These include asphalt, vinyl–asbestos, rubber, and vinyl tile.

An on-grade subfloor is usually a concrete slab poured on the ground over a gravel base, the gravel serving to provide drainage and to replace a clay base, which is subject to swelling and shrinking due to changes in moisture content. The moisture problem is present in an on-grade subfloor, but not to the severe degree present with a subgrade floor.

The resilient floors acceptable below grade are of course acceptable on grade. Cork and backed vinyl tile may also be used on grade.

Any floor that contains an air space below it at least 18 in. high may be termed above-grade. Such a floor could be located below ground level; nevertheless, it is not in direct contact with moisture-transporting material. Moisture migration is not a hazard in such a floor system.

## 20.2  selection of floor

### finishes for wear

There are many ways to classify floor finishes. Two important considerations predominate: wearing characteristics and maintenance. Office and residential floors are subject only to light wear, while some industrial floors receive very severe wear. Such wear is chiefly due to mechanical damage but may be accelerated by chemical attack. The floor of a machine shop collects machining chips, which may embed in the floor; if a lift truck is driven over this very abrasive litter, damage to the floor is greatly accelerated.

For extremely heavy wear, a very hard surface is needed, such as concrete, concrete with a hardener in its surface, hard tile, or in an extreme case, even steel plate. On the other hand, sometimes the best wear resistance is given by a resilient rubber surface which avoids embedment of particles by deforming under them. Concrete which is subject to wear, produces a very dusty surface.

For conditions of light wear the best protection is floor polish. The polish is worn away instead of the floor, which is repolished. Wear of such floors is greatly accelerated by any unevenness due to poor jointing of subfloor boards or lifting of floor tile.

The wear resistance of wood floors depends on the type of wood and the exposed grain. Douglas fir is unsuited to finish flooring because it splinters readily. The end grain of any wood is far superior to other wood surfaces in its wearing qualities.

Other related types of floor damage include indenting and grooving of the floor by heavy loads, and chipping of hard concrete by impact loads.

## 20.3  maintenance characteristics
## of floors

Industrial floors must be swept. Other floors are maintained either by washing or polishing, although frequently either method is possible. Often a carpet is the best choice for maintenance, since carpet cleaning need not be done with the frequency required in washing and polishing operations. If maintenance of a floor may involve also renewal of portions of the floor, then floor finishes consisting of small units such as tiles or blocks are advantageous.

Polishing provides both cleanliness and a bright appearance, but it gives a slippery surface. Certain floor areas may require a nonslip finish, such as entrance areas, the tops of stairs, or homes for elderly people. The number of accidents caused by slippery floors is much larger than is generally realized; in some countries slippery floors appear to account for 50 per cent of accidental deaths. Clearly this floor characteristic receives insufficient attention. Terrazzo floors, especially, are quite slippery and ought not to be polished. Highly polished vinyl tile is only slightly less dangerous. Carpets provide the best footing.

## 20.4  wood flooring

A strong and durable industrial floor is sometimes made with $2 \times 3$s laid with the 3-in. dimension vertical, with horizontal nailing. If the $2 \times 3$s are carefully sawed, such a floor is reasonably level. Its rough surface provides excellent footing and is not tiring to walk on. If operations are such as to deposit dirt or oil on the floor, such a floor is unacceptable, since it cannot readily be cleaned. Such a floor structure is usually limited to mezzanines and storage areas.

A better floor, smoother, cleaner, and more attractive, is the wood block floor. This floor can be given a polish, and is not restricted to industrial floors. The floor blocks are usually oak in small sizes such as $2 \times 2 \times 3\frac{1}{2}$

in., giving a finish floor 2 in. thick. The blocks have an edge-grain face because this face has the best wearing quality and is not subject to splintering. The blocks are usually laid with mastic adhesive on a concrete slab or held by nailing strips (Fig. 20.4).

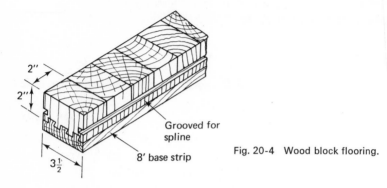

Fig. 20-4   Wood block flooring.

Parquet flooring is a pattern of blocks of hardwood, usually oak. The blocks are laid in such patterns as herringbone, basketweave, rectangular, or square (Fig. 20.5). Parquet flooring is about $\frac{3}{4}$ in. thick in various dimensions from $2\frac{1}{4} \times 6\frac{3}{4}$ to $18 \times 18$ in. A common size is $9 \times 9$ in. The blocks are either nailed or bonded with mastic.

Fig. 20-5   Parquet floor laid in a square pattern.

Strip flooring is made from rigidly graded and kiln-dried hardwoods and softwoods. The wear resistance of wood strip flooring is highest for maple, then birch, quarter-sawed oak, edge-grain yellow pine, edge-grain fir, plain-sawed oak, flat-sawed yellow pine, Norway pine, and least for white pine.

## 20.5   softwood strip flooring

*Softwood strip flooring* is available in standard and heavy types. Standard flooring strip is used for such buildings as schools, offices, and residences. Heavy flooring is specified for industrial floors or special floors such as dance floors and sports arenas. Both types of flooring are made from southern pine and Douglas fir, although a few other species, such as white pine, redwood, and hemlock, may also be used.

Heavy-duty flooring is made in thicknesses of 2, $2\frac{1}{2}$, 3, or 4 in. and widths of 4, 6, 8, 10, or 12 in. These are nominal sizes: a floor strip 2 × 12 would actually measure the usual $1\frac{5}{8}$ × $11\frac{1}{2}$. Edges are either tongue-and-groove, shiplap, or grooved both sides. If grooved both sides, a wood spline is used to join adjacent strips (Fig. 20.6).

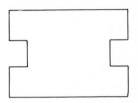

Fig. 20-6 Wood strip flooring grooved for splines.

Most standard softwood strip flooring is $\frac{25}{32}$ in. thick, although other thicknesses are available (Fig. 20.7). Available widths are shown in the figure. Edges are tongue-and-groove, although shiplap and ploughed-for-spline edges are also produced. End-matched boards have tongue-and-groove ends also. The wider dimensions of strip flooring are cupped on the bottom surface to prevent warping.

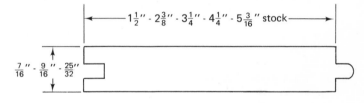

Fig. 20-7 Dimensions of standard softwood strip flooring.

As is usual for wood products, the grading system for wood flooring is complex, and grading rules must be consulted to understand the type and number of defects allowed.

## 20.6 hardwood strip flooring

*Hardwood flooring* is made from oak, hard maple, beech, walnut, and birch, kiln-dried. The usual dimensions and styles are given in Fig. 20.8, although larger sizes are available. The hollow-back strip is the one usually installed. End-matched flooring has tongue-and-groove on all four sides. The back grooves prevent warping due to moisture changes.

Grading rules are, as usual, complex. Prefinished oak flooring is graded Prime, Standard, and so on; other hardwoods are graded First, Second,

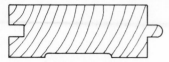 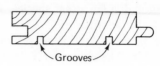

Hollow back **tongue-and-groove** sides and ends $\frac{13}{16}'' \times 2\frac{1}{4}$ tongue-and-groove

Tongue-and-groove sides and ends $\frac{3}{8}''$ or $\frac{1}{2} \times 2''$ tongue-and-groove

Grooves

Flat back $\frac{5}{16}'' \times 2''$ square edge

Fig. 20-8  Common sizes and types of hardwood strip flooring.

Third, Fourth, with some special grades also. Different associations grade different species.

Colonial plank is a special type of installation. Strips are of random widths and have round inserts of a contrasting wood. Tongue-and-groove edges are supplied.

Strip flooring, whether hardwood or softwood, should be laid in the direction in which most wear is likely to occur, for example, running to and away from an entrance door. Occasionally, the direction of the subfloor may prevent this, since the finish floor must be laid at right angles to subfloor boards, unless the subfloor is diagonal to the joists, in which case the finish floor should be laid across the floor joists. Space for expansion of the flooring across the grain must be allowed at the walls of the room.

## 20.7  concrete floors

Light-duty concrete floors should have a minimum compressive strength of 3500 psi, with 3–4 in. of slump. Heavy-duty floors require a stronger concrete of 8000 psi or more with a metallic aggregate or hardener in the surface. Slump should be 1–2 in. so that water will not bleed to the surface to produce a soft surface of cement. Such aggregates as quartz, granite, or traprock are preferred, since it is the aggregate that must sustain such industrial loads as heavy-duty floors receive. Almost all industrial floors must sustain the movement of lift trucks, and these internal vehicles are surprisingly heavy. Some industrial floors are poured as a base slab with a top coat of floor concrete suitably formulated. In the case of floors that must sustain severe abrasion or impact, reinforcing rods must not be so close to the surface that flaking of the concrete will expose them.

The surface of new concrete, even if high-strength concrete is used, is weak. The vibrating and troweling operations of the installation cause the lighter ingredients, cement and water, to rise to the surface. When this lighter material sets, it is called *laitance*. Such a surface is attractively smooth, but is deceptively weak and is not resistant to abrasion. Laitance usually is present to a depth of about 0.05 in. Cement paste has little abrasion resistance, hence the emphasis given to aggregates when designing an abrasion-resistant concrete floor.

Recommended amounts of metallic aggregates (in lb per ft²) applied to the freshly poured surface follow this schedule:

| | |
|---|---|
| for heavy foot traffic | 0.5 |
| light-wheeled traffic and carts | 0.75 |
| heavy-wheeled traffic | 1.25 |
| heavy-wheeled loads | 10 |
| heavy impact | 10 |

Such aggregate is worked into the surface with trowels and floats. Finishing with wooden tools produces a roughened nonskid surface.

Colored concretes are produced by inorganic pigments added by shaking over the surface and floating the material into the surface. Some of these coloring additives include silica sand. The end result of coloring a concrete may not be entirely satisfactory if installed by an unskilled or inexperienced crew.

Concrete emulsions are sometimes employed as finish floors over concrete subfloors. Cement may be mixed with synthetic resins or a rubber latex with various fillers, such as sawdust and pigments. These are applied as finish layers about $\frac{1}{4}$ in. thick.

A maximum variation in level of plus or minus $\frac{1}{8}$ in. for every 10 ft in any direction should be the maximum out-of-level allowance for a concrete floor. If drains are installed, the floor must have a gentle slope toward them.

### 20.8  monolithic floor surfaces

A monolithic structure is one without joints. A concrete floor is an example. Other types of monolithic floor include terrazzo, mastic, and magnesite floors. In recent years a considerable number of new kinds have been developed. Such floors may have a remarkable variety of names: monlithic surfaces, poured floors, formed-in-place floors, toppings, seamless floors, or troweled composition floors. All these floors are installed by mixing together at the jobsite a binding material or matrix with fillers and decorative additives. The binder hardens in place to form a seamless floor. Such floors are finished either by self-leveling, by troweling, or by surface grinding. Installation must usually be done by specialist crews trained in the required techniques.

A poured-in-place floor may be installed over any subfloor that is sound, that is, without cracks or voids. If such openings are present, they must be suitably patched or sealed.

## 20.9   asphalt seamless floors

The frequent reference to asphalt in this book suggests the great versatility of this material in construction. As a flooring material, asphalt is used both in floor tile and for seamless floors. However, asphalt flooring is used less nowadays than it was in the past.

Asphalt membrane floors are made by saturating a thick felt with asphalt. Since a floor must carry concentrated loads, the saturating asphalt used for a floor must be harder than a roofing asphalt.

The asphalt mastic floor is made with a mix of asphalt emulsion, portland cement, sand, and gravel or crushed stone. This is spread over the subfloor, leveled, and compacted. The approximate composition of such floors is given in parts by volume in the following table:

| | Light-Duty Floor: $\frac{1}{8}$–$\frac{3}{4}$-in. Thick | Heavy-Duty Floor: $\frac{1}{2}$–$1\frac{1}{2}$-in. Thick |
|---|---|---|
| asphalt | 1.5 | 1.5 |
| cement | 1.0 | 1.0 |
| sand | 3.5 | 2.0 |
| stone or gravel | — | 3.5 |
| water | sufficient water to produce a free-flowing and workable mix | |

A primer coat of asphalt is applied to the subfloor.

Since asphalt is thermoplastic, it must not be installed in hot areas where it could soften and flow under the pressure of wheeled traffic. The asphalt floor is not slippery, is able to resist abrasion by deforming, and is not seriously attacked by chemicals. Often asphalt is used to resurface a deteriorated concrete or other floor surface.

## 20.10   terrazzo and magnesite

## floors

*Terrazzo* is a dense and hard concrete finish flooring mottled with marble or other chips laid on a concrete subfloor and finished by grinding. The subfloor is first poured, then a layer of cement and sand is spread on the subfloor as a base for the terrazzo finish. A gridwork of metal or plastic dividing strips is bedded into the sand–cement layer and leveled to about 1 in. above the sand–cement layer (Fig. 20.9). These grid strips are used to divide the floor into the pattern seen in Fig. 20.10. They also serve to eliminate cracking.

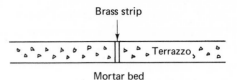

Brass strip

Terrazzo

Mortar bed

Fig. 20-9 Section through a terrazzo floor.

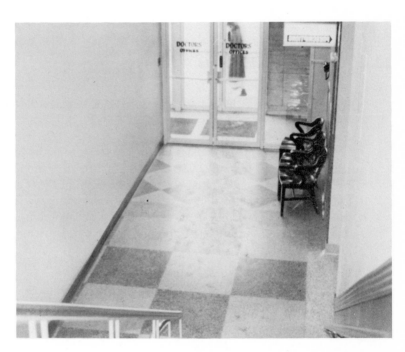

Fig. 20-10 Terrazzo floor in a medical clinic.

Terrazzo topping is proportioned 200 lb of marble or granite chips to 100 lb of portland cement. This gives a somewhat slippery surface. A nonslip heavy-duty terrazzo may have 150 lb of marble chips to 100 lb of cement and 50 lb of abrasive grain. The terrazzo topping is placed in the areas between dividing strips and compacted with heavy rollers to extract excess water, then troweled until the tops of the dividing strips are visible. The surface is kept moist for 6 days, then ground and polish-ground.

Mosaic is a variant of terrazzo. Mosaic combines small pieces of stone, glass, marble, metal, pottery, or other materials to create a design in a bed of cement.

In a *magnesite* floor finish a special cement is used, called magnesium oxychloride cement or Sorel cement. This is a hydrated cement that sets rapidly after mixing with water. The magnesite floor is composed of Sorel cement, fillers such as sawdust, and aggregates such as stone chips. When hardened, this mix produces a smooth, dense surface that is relatively flexible compared to any other ceramic type of floor. If formulated with marble, the floor resembles terrazzo in appearance. As with terrazzo, a special abrasive grain such as aluminum oxide may be added to reduce slipperiness. Heavy aggregates are used for surfaces subject to heavy traffic loads. An unusual weakness of the magnesite floor is that it is not completely water-resistant.

## 20.11　seamless resin floors

The synthetic resins have invaded the field of seamless flooring in recent years. These new types may be based on epoxies, polyesters, silicones, synthetic rubbers, or polyurethanes.

Both polyesters and epoxies are used in terrazzo floors in thinner layers than are used for standard terrazzo floors. As little as $\frac{1}{16}$ in. thickness may be spread. The synthetic material replaces the cement of the standard terrazzo. The chief advantage of the epoxy and polyester terrazzos is their reduced weight. Divider strips are used, and grinding and polishing are performed after 16–48 hr of cure time. If grinding is delayed, it becomes increasingly difficult to perform. Plasticizers are sometimes incorporated in the resin, but given the tendency of plasticizers to migrate and leach from plastics, they probably should not be recommended. Opinion is divided on the subject of plasticizers in flooring applications.

Epoxies are also used in flexible terrazzo-like floor tile.

Polyurethanes for seamless resin floors may be of three types:

1. Single-component moisture-cured by moisture in the air.
2. Single-component oil-modified.
3. Two-component catalyst-cured.

The polyurethane resins produce a superior floor with the properties of abrasion resistance, resistance to impact, flexibility, resistance to chemical attack, and the advantage that they can be installed over existing floors even if these are deteriorated. They bond well to virtually any material, although not to polyethylene film. They are too elastomeric to be used as terrazzo floors, however, that is, too flexible to permit grinding of the floor. Hence they are restricted to self-leveling types of finish floor.

Floors of strikingly attractive appearance may be devised by use of polyurethane. As an example, one proprietary type of floor (Beauty-Flo, by Diasyde Corp.) uses a white base coat of diatomaceous earth filler. Paper-thin color chips (not stone) are then rolled into this base coat to provide a color pattern. The glaze coat or top coat is a clear moisture-cured polyurethane. The resin floor has a nonskid surface and requires no waxing.

The polyester, epoxy, and polyurethane resin floors are weather-resistant and therefore suited to exterior as well as interior applications.

### 20.12 *ceramic floor tile*

### *and brick*

Clay floor tile are produced by the same processes as are used in the manu-facture of other fired clay products, such as brick and tile, and both glazed and unglazed tile are available. The unglazed variety is preferred in flooring. Either type has a hard, nonporous surface, and may be used either indoors or outdoors without damage by frost action. Three types are favored for floors, although other types are also in use: ceramic mosaic tile, quarry tile, and paver tile.

*Ceramic mosaic tile* is made in small sizes, many shapes, and a range of patterns (Fig. 20.11), glazed and unglazed. The unglazed variety is offered in a wide range of colors, but glazed tile is capable of the widest range of color. Nominal thickness is $\frac{1}{4}$ in. Mosaic tile is popular for floors because of its interesting possibilities for design, its durability and hardness, and its convenient maintenance. Such tile can be installed on almost any surface that is rigid and sound. They are supplied in a sheet of tile, using a paper or a rubber backing, $9 \times 9$ in. or larger.

*Quarry tile and pavers* have larger dimensions than mosaic tile. Quarry tile is unglazed, and colors are restricted to the burned colors of red, buff,

Fig. 20-11   Typical ceramic mosaic patterns.

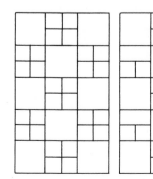
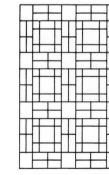

and brown. Shapes are square or oblong: $2\frac{3}{4}$, 4, 6, 8, and 9 in. square, and oblong tile in sizes from $2\frac{3}{4} \times 6$ through $3\frac{3}{4} \times 8$, $4 \times 8$, to $6 \times 9$ in. The most popular size is $6 \times 6$. Quarry tile is $\frac{1}{2}$ to $1\frac{1}{2}$ in. thick. Paver tile, similar to quarry tile but intended for less demanding service, has a thickness of $\frac{3}{8} - \frac{5}{8}$ in. In addition, trim tile is made in cove, straight base, and other shapes. Both quarry and paver tiles have a slightly irregular surface, to provide a nonskid floor.

Ceramic floor tiles are bonded to their base floor by standard cement mortars, dry-set mortars, cement–latex adhesives, or organic adhesives such as epoxy or solvent-release rubbers. Dry-set mortars are made of portland cement with additives that give the mix a high viscosity to retard absorption of the water in the mix by porous substrates or the tile itself. Such mortars are used for bonding tile to concrete block, brick, or concrete.

Brick floors are often used in interior and exterior locations, and paving brick may be seen on the roads of many towns and cities in the eastern and southern states. Brick offers color, pattern variety, durability, and abrasion resistance. A low-absorption, dense type of brick is preferred for ease of maintenance and resistance to staining. Floor brick may be laid with or without mortar (Fig. 20.12). It is usual to lay brick over a concrete slab for interior floors, although exterior brick walks and patios may be laid on gravel.

Fig. 20-12 Brick paving in shopping center at Reston, Virginia.

The group of *organic floor tiles* includes linoleum tile, vinyl, vinyl asbestos, asphalt, cork, and rubber tile, with new types recently introduced such as epoxy, polyester, polyurethane, and even Hypalon tile. The standard size for these tiles is $9 \times 9$ in., with larger sizes also in use. The available thicknesses differ from one type to another. The wide range of tile materials requires that a bonding adhesive be selected that is suited to the tile to be laid.

A tile floor can be no smoother than the subfloor. All dust, dirt, oil, grease, and moisture must be removed from the subfloor before the tile is laid, and concrete must be thoroughly cured.

## 20.14 cork tile

*Cork tiles* are made from the outer bark of the cork oak tree found in southern Europe. The cork is made into two types of flooring: cork carpet sheets and cork tile. Cork carpet is obtainable only on special order, but cork tile is a standard flooring. The tiles are made by blending phenolic or other thermosetting resin with cork granules and baking under heat and pressure to produce a strong tile of low porosity. Available colors are three shades of brown, produced by different baking methods.

Cork gives the most resilient of all floor surfaces, but also the most porous, except of course for carpet. It is also an excellent nonskid surface. It is unsuited to any type of heavy-duty service. Thicknesses of $\frac{1}{2}$, $\frac{3}{16}$, and $\frac{5}{16}$ in. are produced. Four finishes are offered by manufacturers: natural, waxed, resin-reinforced waxed, and vinyl. The natural finish has an untreated surface, while the waxed cork is impregnated with a floor wax. The resin-reinforced surface uses a resin-reinforced wax. The vinyl cork tile has a wear layer of vinyl, polycarbonate, or polyurethane. This last type best resists staining and must not be sanded in the course of installation, as is done with other types of cork tile.

Cork tile is not suitable for floors below grade.

## 20.15 asphalt tile

*Asphalt tiles* were one of the earliest types of organic tile to be developed, gaining rapid acceptance because they were at the time of development the only resilient floor tile that could be used on-grade or below grade. Although they have lost markets to other types of organic tile, they are the least expensive of the resilient tile.

Formerly, asphalt tiles were composed of about 50 per cent asbestos fiber, 25 per cent mineral filler, and 25 per cent gilsonite asphalt by weight. Such tiles are still produced, but only dark colors are possible. Light-colored and marbled-colored asphalt tiles are also produced now. These contain little asphalt, their composition being about one-third cumarone–indene resin binder with asphalt, 60 per cent asbestos fiber, and limestone filler with inorganic pigments.

Asphalt tiles are manufactured in $\frac{1}{8}$- and $\frac{3}{16}$-in. thicknesses. They may be laid over any sound concrete, wood, or asphalt mastic base, using an asphaltic adhesive. A smooth subsurface is required, since asphalt tile is somewhat brittle and can be cracked on an uneven surface.

Asphalt tiles will not support combustion and are resistant to cigarette burns. They are somewhat noisy and more difficult to clean and polish than other tile. Most asphalt tile is applied to commercial and industrial floors.

## 20.16   vinyl asbestos

## and vinyl tile

*Vinyl asbestos tiles* are suitable for subgrade floors and have tended to replace asphalt tile in recent years. In this tile, polyvinyl chloride, polyvinyl acetate, and plasticizers replace the asphalt or cumarone-indene used in asphalt tile. Vinyl asbestos tile is restricted to indoor installations but may be used below grade on concrete subfloors. It has limited resistance to indentation but is not seriously harmed by cigarette burns, except for color damage. Thicknesses are $\frac{1}{16}$, $\frac{3}{32}$, and $\frac{1}{8}$ in.

Vinyl tiles are usually made of polyvinyl chloride without the polyvinyl acetate also used in vinyl asbestos tile. Vinyl tile is somewhat more expensive, but lacking the gray color effect of asbestos, it is more attractive in appearance.

The cost of vinyl tile depends on the amount of vinyl in the tile. Cheaper grades may contain 35 per cent PVC to 65 per cent filler, whereas the high-priced translucent grades may contain almost 100 per cent PVC binder. Five thicknesses are produced: 0.050, $\frac{1}{16}$, 0.080, $\frac{3}{32}$, and $\frac{1}{8}$ in. A greater variety of patterns and colors is offered than is available in any other type of floor tile.

Vinyl tile can be installed on virtually any floor, regardless of grade. As is necessary before laying any finish floor, a concrete subfloor must be fully cured before vinyl tiles are laid.

Backed vinyl tile is made of a cheaper backing material for the bulk thickness of the tile, with a wear layer on top with a thickness from a few thousandths of an inch to 0.050 in. in an expensive grade. The backup material may be a cheap grade of vinyl or various nonvinyl compositions.

*Rubber tiles* are made from SBR rubber, styrene–butadiene, which is the rubber used in automobile tires. This has better aging properties than natural rubber and its oil resistance qualifies it for a flooring material. Thicknesses of 0.080, $\frac{3}{32}$, $\frac{1}{8}$, and $\frac{3}{16}$ in. are produced. Such rubber flooring is also available in 36-in. rolls. These rubber flooring products may be used below grade. Rubber tile is declining in popularity, however.

## 20.17   linoleum

*Linoleum* is the oldest type of resilient finish floor. It is usually applied as sheet goods but is also available as tile. The decorative surface consists of oxidized oils such as linseed oil and various resins, with fillers and pigments. This surface material is applied to a felt carrier, which may be a mineral or organic fiber such as burlap. The drying oils are thermosetting. Since these oils have a yellow or yellow-brown color, white and pastel shades are not easy to produce.

Standard-gage linoleum has a minimum wear layer of 0.050 and a total thickness of 0.090 in; heavy-duty linoleum is 0.125 in. thick and has been called battleship linoleum, since it was first manufactured for warship service. Linoleums are not installed below grade because of sensitivity to moisture.

## 20.18   carpeting

Originally *carpets* were manufactured from natural fibers. Wool was used for quality carpets and cotton for low-priced rugs. Wool now competes against a constantly growing number of synthetic fibers, the latter now being used in three-fourths of all carpets manufactured. Domestic sheep produce too fine a wool for carpeting, so all carpet wools are imported. Wool was surpassed by nylon about 10 years ago, and acrylic, polypropylene, acetate, and other synthetics also compete in the carpet market.

A carpet fiber must have resilience, which is the ability to recover quickly from traffic pressure, together with resistance to wear, ability to be colored, rot resistance, some degree of fire resistance, and stain resistance. It is also desirable for the fiber to resist the generation of electrostatic charges. Polypropylene is almost free of electrostatic charge buildup, but its resilience does not compare with that of wool or nylon.

The grid of yarns in a carpet contains warp yarn and weft yarn. The warp yarn refers to the backing yarns that run lengthwise in the carpet. These are of two types, both shown in Fig. 20.13. The chain yarns bind the carpet material together by weaving alternately over and under the weft yarn to pull down the pile yarn tight to the stuffer yarn. The stuffer warp yarns are straight, as indicated in the figure. The weft yarn comprises the backing yarns that run crosswise of the carpet and about which the pile yarn are hooked.

Carpet quality is most closely related to pile density, the weight of pile yarn per square yard of carpet. A denser pile will have more pile tufts per square inch of carpet.

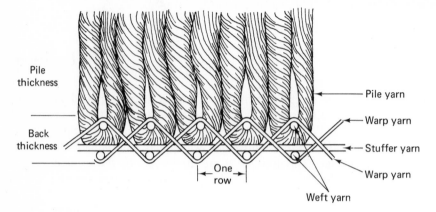

Fig. 20-13  Weave of a carpet.

## QUESTIONS

**1** What characteristics are generally required (a) in an industrial floor; (b) in an office floor; (c) in the floor of a prestige office?

**2** What advantages are offered by a modular floor (tile floor) over sheet flooring?

**3** Enquire about installation and maintenance costs for floors of commercial buildings. Which is the larger cost, the cost of installation or maintenance costs over 5 years?

**4** How does the position of the subfloor with respect to grade influence the selection of a finish floor material?

**5** (a) What types of floor tile may be installed below grade? (b) What types of floor tile may not be so installed?

**6** (a) List areas in buildings where a slippery floor would be a serious hazard. (b) What floor materials could be installed in such areas?

**7** Why is wood strip flooring grooved at the back?

**8** How do you select the direction in which wood strip flooring should run?

**9** What is the difference between a terrazzo and a magnesite floor?

**10** What synthetic resins are used in seamless floors?

**11** What is the difference between quarry and paver floor tile?

**12** State the advantages and the disadvantages of cork flooring.

**13** How would you define linoleum as a material?

**14** List the desirable characteristics of a carpet fiber.

**APPENDICES**

# DESIGN
# PROBLEM

*Appendix*

In the far northern regions of Siberia, Canada, and Alaska, the soil is permanently frozen and so is called *permafrost*. It is not possible to carry utilities such as sewers and water in the ground because the pipes would freeze. Instead, insulated trenches carried on grade or above grade are used to protect utilities distributed to buildings in a town. These trenches are called *utilidors*. A utilidor is shown under construction at an army base in Fig. A.1. This one is made of concrete insulated on the inside and backfilled with gravel. The utilidor is covered over with wood platforms about 3 ft long, made of 2 × 3s nailed together and carried on light angle-iron stringers. These covers are removable for maintenance of the utilidor. A separate steam tracer line keeps the utilidor from freezing.

Fig. A-1  Utilidor in a permafrost area.

The North is waiting for a good utilidor design. It is very expensive to ship materials north, and the weight of materials for the utilidor is a significant factor. It is also expensive to maintain skilled construction labor in these northern regions, so a prefabricated utilidor design would be desirable, if possible. Maintenance is a serious problem: the interior of a utilidor on a cold day might reach temperatures of +140°F; temperature control is impractical, and much water collects in the utilidor, chiefly due to melting of snow and condensation of leaking steam. 1. Make a cross-sectional layout of the interior of a utilidor, showing an 8-in. sewer line, water line, building steam heat line, steam tracer (1 in.), electrical power line, and telephone line. The sewer line must be at a lower elevation than the water line. Use plastic pipe if you wish, except for steam. This layout decides the required width and height of your utilidor. 2. Design the utilidor and select construction materials for it. Use no materials that absorb, or are deteriorated by, water. The cross section could be rectangular, circular, triangular, or other. Outside design tmperature is −60°F; however, heat loss (steam consumption) is not a critical matter.

# NOTE
# ON METRIC UNITS

*Appendix*

The basic SI (metric) units are:

| | |
|---|---|
| mass | kilogram (kg) |
| length | meter (m) |
| time | seconds (s) |
| force | newtons (N) |

One newton is the force required to give a mass of 1 kilogram an acceleration of 1 meter per second each second. One pound force = 4.448 N.

Stress is expressed in newtons per square meter ($N/m^2$). One $N/m^2$ = 1 pascal (1 Pa). One pound per square inch = 6894.8 $N/m^2$ = 6894.8 pascals.

$$1 \text{ kilogram} = 2.2 \text{ pounds}$$

$$1 \text{ meter} = 39.37 \text{ inches}$$

$$1 \text{ inch} = 2.54 \text{ centimeters}$$

**GLOSSARY**

*Appendix* **C**

**absolute volume.** Volume of the solid particles of a loose granular material

**absorptivity.** Relative ability to absorb sound or light

**acoustics.** Science of sound

**aerosol.** Dispersion of a liquid in a gas

**aggregate.** Loose granular filler material added to cement to make concrete

**air-entraining agent.** Material that introduces small air bubbles into concrete

**ambient temperature.** Temperature of the atmosphere

**anchor bolt.** Bolt embedded in foundation concrete for anchoring a superstructure

**anodizing.** Development of an aluminum oxide protective surface on aluminum by an electrical bath process

**asphalt.** Semisolid mixture of hydrocarbons derived from petroleum


**backfill.** Earth filled in around a foundation wall to replace earth removed during excavation

**backup wall.** Wall supporting a facing material

**balloon framing.** Wood frame construction in which the studs extend in one piece from the foundation sill to the roof plate

**base sheet.** Saturated or coated felt used as the first ply in a built-up roof

**bat.** Half a brick, produced by breaking off one end of the brick

**batt.** Semirigid block of mineral wool insulation

**batten.** Narrow strip of wood nailed over a wood joint

**beam pocket.** Recess at the top of a foundation to receive the end of a floor beam or joist

**bearing wall.** Wall that supports all or part of the roof or floors of a building

**belt course.** Course of stone in a masonry wall

**bitumen.** Semisolid mixtures of hydrocarbons from petroleum or coal

**bleeding.** Migration of a volatile liquid to the surface of a liquid mixture

**board foot.** Volume of material measuring 1 ft $\times$ 1 ft $\times$ 1 in.

**bond beam.** Beam that ties together an upper and a lower section of a wall to introduce additional rigidity

**bottom chord.** Lower main horizontal member of a truss

**Btu (British thermal unit).** Amount of heat required to raise the temperature of 1 lb of water from 60 to 61°F.

**built-up roof.** Roofing built up of successive felt plies bonded with bitumen


**calcine.** To harden clay materials at a high temperature

**cant strip.** Molding of triangular cross section with a 45-degree slope, fitted into the angle between a structural deck and a roof

**cap sheet.** Mineral-surfaced coated felt used as the top ply of a built-up roofing membrane

**catalyst.** Compound added to a synthetic resin to harden (cure) it

**caulking compound.** Sealing material of limited ability to elongate

**cavity wall.** Masonry wall of two thicknesses of material separated by a 2-in.-wide cavity

**cement.** Adhesive

**ceramics.** Materials from the rock and soil of the earth's crust

**chalking.** Type of paint failure distinguished by the appearance of loose powdery pigment at the surface

**cladding.** External covering over a base material

**clear lumber.** Lumber free of knots and blemishes

**closed cells.** Noninterconnected cells that are separated by membranes of material

**coal-tar pitch.** Semisolid hydrocarbon produced from the distillation products of coal

**common brick.** General-purpose brick not specially manufactured for resistance to weathering and moisture penetration

**consistency.** Degree of softness or firmness of a compound

**cornice.** Small projection at the top of a wall

**corrugated.** Given a curved pattern of symmetrical hills and valleys

**course.** Horizontal tier of brick, tile, or block

**coverage.** Hiding power of a paint, expressed in square feet per gallon

**creep.** Continuous deformation with time

**cure.** To harden by chemical action

**curing agent.** One part of a two-part compound, which, when added to the other part, will set up the curing (hardening) action; also called a **catalyst**

**curtain wall.** Wall of lightweight construction supporting only its own weight

**cutback.** Solvent-thinned cold-process bitumen

**dado.** Rectangular groove cut across the grain of the wood

**decibel.** Measure of the power of a sound

**dew point.** Temperature at which water vapor begins to condense in cooling air

**dress.** To plane one or more sides of a piece of sawed lumber

**drywall construction.** Interior cladding with panels of board, using no wet plaster

**ductile.** Capable of considerable stretch or elongation

**eave.** Lower part of a roof, projecting beyond the walls

**efflorescence.** White deposit on the surface of masonry or concrete, consisting of hydrated salts that moisture has transported to the surface of the masonry

**elastic.** Capable of deformation with recovery of original dimensions when the deforming stress is removed

**elastic limit.** Highest stress at which a material behaves elastically

**elastomer.** Rubber, characterized by remarkable elastic deformation

**emboss.** To raise portions of a surface in a decorative effect

**emulsion.** Dispersion of globules of one liquid in another liquid, the two liquids not being soluble one in the other

**E value.** Ratio of stress to strain; rigidity of a material when stressed

**exotherm.** Heat released during curing of a material

**extruded.** Shaped by forcing through a contoured die opening

**face brick.** Brick of superior appearance and weather resistance for exterior applications

**factor of safety.** Ratio of the strength of a material to its allowable design strength

**fascia.** Board nailed to the ends of the roof rafters around the eaves of a building

**felt.** Fabric of interlocking fibers made without weaving

**firewall.** Fireproof interior wall to confine a fire

**flashing.** Waterproofing seal at the edge of an impermeable roofing membrane

**flash point.** Temperature at which a flammable material will first flash into flame

**flexure.** Bending

**float.** Wooden trowel

**flood coat.** Top coat of bitumen in a built-up roof with a surfacing aggregate

**flux.** Material added to dissolve and lower the melting point of an oxide or other ceramic material

**footing.** Lowest part of a foundation, a concrete slab wider than the foundation wall

**frog.** Depression, as in the large face of a brick

**furring.** Building out or leveling of a surface as a base for attachment of other materials or cladding

**gable.** End wall of a building which comes to a triangular point under a sloping roof

**gage.** Measure of thickness for metal sheet and wire; higher gages are thinner than lower gage numbers

**galvanic corrosion.** Electrical type of corrosion in which two dissimilar metals and a liquid form a natural electric battery; only one of the two metals of this battery is corroded

**gasket.** Preformed shape used to seal a joint

**glaze.** Fused vitreous, hard, and smooth surface finish for clay products

**glulam.** Glued laminated timber

**grade.** Slope of the terrain or the level of the terrain at the jobsite

**grain.** Measure of weight for water vapor; 7000 grains are equal to 1 pound

**grout.** Nonshrinking concrete filler and sealer material

**gun consistency.** Degree of consistency required to apply a semisolid material through the nozzle of a caulking gun

**haydite.** Calcined bloating clay used as aggregate

**head.** Topmost horizontal member in window sash

**hydration.** Chemical curing or hardening of cement by the action of water

**hydraulic cement.** Cement that hardens with water

**hydrocarbon.** Chemical compound of carbon and hydrogen, such as gasoline or asphalt

**initial set.** Original stiffening of a cement paste

**jamb.** Vertical member at the side of a window or door sash

**kerf.** Groove cut in wood while it is being sawed

**kiln.** Drying or calcining oven or furnace

**knife consistency.** Degree of consistency required to apply semisolid material with a putty knife

**kraft paper.** Strong grade of paper, usually brown in color, made from sulfite pulp

**lag screw (bolt).** Threaded wood screw with a wrench head

**laitance.** Soft and weak layer of cement and water that rises to the surface of a concrete pour

**laminate.** Material made up of several layers

**latex.** Emulsion or dispersion of a liquid in another liquid, the two liquids not being mutually soluble

**lath.** Base for attachment of plaster

**light.** Pane of glass

**lintel.** Beam spanning a wall opening

**mastic.** Compound that remains pliable with age

**mesh.** Number of openings per lineal inch in a sieve

**mil.** Thousandth of an inch

**mild steel.** Steel alloy with a low carbon content, 0.2 per cent carbon or less

**modular size.** Dimension that conforms to a given module, such as 48-in. sheets for studs 16 in. on centers

**module.** Standard dimension to which units conform, such as studs 16 in. on centers

**modulus of elasticity.** Ratio of stress to strain

**modulus of rupture.** Tensile stress at failure in a bending test performed on brittle materials

**monolithic.** Continuous structure without joints

**mopping.** Application of hot bitumen

**mortar.** Mixture of materials used to bond brick, block, or tile

**mullion.** Vertical member that holds together two adjacent units of sash or curtain wall

**neat.** Without additives; neat cement has no aggregate

**organic.** Having a chemistry based on carbon, such as wood

**parging.** Coat of plaster or cement mixture used to smooth a masonry wall

**party wall.** Wall that is common to two buildings and is the boundary between them

**patina.** Oxide film formed on copper by weathering

**penetration.** Depth to which the metal of a plate is melted during welding

**perimeter insulation.** Insulation installed around the outer walls or foundation of a building

**perm.** Unit expressing the rate of penetration of water vapor through a material

**permeability.** Rate of passage or migration of a liquid, usually water, through a porous material

**pilaster.** Built-in masonry column in a masonry wall

**pitch.** Slope of a roof

**pitch pocket.** Metal container filled with bitumen or flashing cement to seal around an element penetrating a roof

**plasticity.** Ability to be deformed into a different shape or dimension

**plasticizer.** Liquid material added to certain plastics to reduce their hardness and increase pliability

**ply.** Layer in a series of layers

**polymer.** Chemical compound built up by a chain of simple chemical modular units (monomers)

**porcelain enamel.** Fused vitreous smooth surface baked on metal sheet

**pozzolan.** Silicate material used in formulating hydraulic cements

**prefabricated.** Partially fabricated before delivery to the jobsite

**primer.** First coat of a surface finish applied to improve the adhesion of the following coats

**proportional limit.** Highest stress for which there is a linear relationship between stress and strain

**purlin.** Roof structural member spanning the space between roof trusses and carried by the trusses

**quarter-sawed.** Sawed so as to produce edge-grain lumber

**rabbet.** Groove in wood parallel to the grain
**racking.** Tendency of a rectangular frame to distort from its rectangular shape due to lack of stiffness against shear forces
**rafter.** Beam supporting a roof
**rebar.** Steel bar for reinforcing concrete
**reflectivity.** Relative ability of a surface to reflect light or sound
**resilient.** Elastic, having the ability to regain an initial shape after deformation
**resin.** Organic synthetic chemical
**ridge roll.** Special roofing shape to fit over the ridge of a roof

**sagging.** Running of liquid compounds when applied to a vertical surface
**sandwich panel.** Panel made by facing a soft material with harder and stiffer skin facings
**sash.** Frame of a window in which the glass is set
**sealant.** Mastic material used to seal joints
**sealer.** Material to seal a surface against moisture penetration
**shake.** Granular material applied to a floor by sprinkling
**sheathing (sheeting).** Lumber that covers a framework of studs
**sill.** Horizontal base of a frame
**sizing.** Primer coat used to fill the pores of a porous material to render it less porous for the following coats
**skinning.** Drying of the surface of a coating while the interior of the compound is still moist
**slag.** Waste material from the process that refines a metal from its ore
**slaking.** Adding water to hot lime
**slurry.** Heavily watered and pourable mixture of water and some powdered material
**soffit.** Underface of a floor or an eave
**solar screen.** Masonry wall pierced with openings, intended to serve as a screen against sunlight or noise
**sole plate.** Bottom plate in wood framing
**span.** Distance between supports or columns carrying a beam
**spandrel beam.** Beam from column to column in outside walls, carrying the curtain-wall panels above it
**specific gravity.** Ratio of the weight of 1 cubic foot of a material to the weight of 1 cubic foot of cold water
**square.** 100 square feet of a roofing material
**steel.** Alloy of iron with carbon, silicon, and manganese
**story.** Part of a building between two adjacent floors
**strain.** Deformation produced by a stress
**stress.** Load per square inch of the area supporting the load
**striated.** Marked with small parallel grooves
**stud.** Vertical members in a wall frame
**subfloor.** Structural floor supporting the finish floor material
**substrate.** Subsurface to which a surface film or material is attached
**suction.** Ability of a masonry surface to absorb water
**synthetic.** Artificial or manufactured, not obtainable naturally

**terra cotta.** Decorative ceramic veneer

**thermal conductivity.** Rate at which heat is conducted through a material

**thermoplastic.** Softenable by the application of sufficient heat

**thermosetting.** Not softenable by heating after curing

**toenail.** Nail driven diagonally through a piece of lumber

**top chord.** Upper main horizontal member of a truss

**translucent.** Transmitting diffused light but not transparent

**transparent.** Transmitting light almost completely

**truss.** Braced frame, usually including diagonal bracing

**tube.** Hollow bar

**ultimate tensile strength.** Maximum tensile stress exhibited by a material up to the point of rupture

**united inches.** Sum of the length and width of a light of glass

**vapor barrier.** Film barrier to the passage of water vapor

**vehicle.** Liquid portion of a mixture such as paint

**veneer.** Thin sheet of a material for attachment to another surface

**viscosity.** Resistance of a liquid to flow or shear forces

**vitrification.** Changing of a ceramic material to a glassy material at an elevated temperature

**volatile.** Having the characteristic of evaporating readily at room temperature

**weathering.** Aging process resulting from atmospheric exposure

**wythe.** Brick wall on either side of a cavity in the wall

**yard lumber.** Lumber for building construction

**yield stress.** Stress at which a material begins to deform plastically

**Young's modulus.** Ratio of stress to strain, same as modulus of elasticity

**INDEX**

Manganese, 156, 160
Marble, 50
Mastics, 266
Masonry cement, 79
Mass law for sound, 40
Medullary rays, 196
Metals, 3
Metamorphic rock, 48
Mineral wool, 151, 322
Modulus of elasticity, 15
Modulus of rupture, 16
Moh's scale of hardness, 48
Mortar, 109, 133
Mosaic floor, 365
Movement, 8

**N** Nails, 181, 210
Nickel, 160
Noise reduction coefficient, 42

**O** Octane, 229
Olefins, 231
Open-web steel joists, 177
Organic materials, 4, 6
Oxides, 55
Oxygen converter, 154

**P** Paint, 281
Paraffins, 231
Particle board, 221
Partition tile, 127
Peel strength, 280
Perlite, 85, 321
Perm, 30
Phenol-formaldehyde, 225, 227, 243, 244
Pigment, of paint, 282
Pilaster block, 114
Pipe, 169
  corrugated, 170
Plaster, 140, 350
Plasticity, 12
  of soils, 63
Plasticity index, 64

Plastic limit, of soils, 63
Plate glass, 347
Plywood, 213
Polycarbonate, 225, 237
Polyesters, 245
Polyethylene, 227, 233, 237
  foam, 254
Polymers, 223
Polymethyl methacrylate, 237
Polypropylene, 233
Polystyrene, 227
  foam, 255, 256, 294, 296, 324
Polysulfide rubber, 250, 270, 301
Polyurethane, 13, 225, 250
  foam, 258, 294, 300, 321, 323
  floor, 366
Polyvinyl chloride, 227, 234, 236
  foam, 255
  siding, 317
Polyvinyl fluoride, 235, 237
Porosity, of rock, 49
Pozzolans, 80
Pressure-sensitive adhesive, 280
Proportional limit, 14
Putty, 269

**Q** Quoin, 52

**R** Red lead, 284
Reflectivity, 36
Reinforced plastics, 245
Reinforcing bar, 171
  mesh, 172
Retarders, 95
Retentivity, of mortar, 133
R-factor, 27
Rimmed steel, 160
Rockwell hardness, 185
Roll roofing, 219, 301
Roman brick, 121
Roofing felts, 219, 298
Roofing sheet, 175, 305
  tile, 304
Rubber tile, 371
Rubble stone, 51
Running bond, 111, 124
Rutile, 57